# THE TYRANNOSAUR CHRONICLES

**Also available in the Bloomsbury Sigma series:**

*Sex on Earth* by Jules Howard
*p53: The Gene that Cracked the Cancer Code* by Sue Armstrong
*Atoms Under the Floorboards* by Chris Woodford
*Spirals in Time* by Helen Scales
*Chilled* by Tom Jackson
*A is for Arsenic* by Kathryn Harkup
*Breaking the Chains of Gravity* by Amy Shira Teitel
*Suspicious Minds* by Rob Brotherton
*Herding Hemingway's Cats* by Kat Arney
*Electronic Dreams* by Tom Lean
*Sorting the Beef from the Bull* by Richard Evershed
and Nicola Temple
*Death on Earth* by Jules Howard

# THE TYRANNOSAUR CHRONICLES

## THE BIOLOGY OF THE TYRANT DINOSAURS

David Hone

BLOOMSBURY
sigma

Bloomsbury Sigma
An imprint of Bloomsbury Publishing Plc

50 Bedford Square              1385 Broadway
London                        New York
WC1B 3DP                      NY 10018
UK                            USA

www.bloomsbury.com

BLOOMSBURY and the Diana logo are trademarks of
Bloomsbury Publishing Plc

First published 2016

Photo credits (t = top, b = bottom, l = left, r = right, c = centre)
Colour section: P. 1: Carnegie Museum of Natural History / D. Hone (t); Los Angeles County
Museum / D. Hone / A. Farke (b). P. 2: New Mexico Museum of Natural History and Science (t);
X. Xu (tr, cr, br); J-C. Lü (b). P. 3: X. Xu (t); D. Hone (cl); J. Mallon / Royal Ontario Museum
(cr, b). P. 4: D. Hone / Royal Tyrrell Museum of Palaeontology (t); P. Currie (b). P. 5: Mongolian
Academic of Sciences / Hayashibara Museum of Natural Sciences-Mongolian Paleontological
Centre joint expedition (t); D. Hone / Dinosaur Isle Museum (cl); P. Falkingham (cr); The Royal
Saskatchewan Museum (b). P. 6: Mongolian Academic of Sciences / Hayashibara Museum of Natural
Sciences-Mongolian Paleontological Centre joint expedition (t); D. Hone / Mongolian Academic
of Sciences / Hayashibara Museum of Natural Sciences-Mongolian Paleontological Centre joint
expedition (c); L. Witmer / Mongolian Academic of Sciences / Hayashibara Museum of Natural
Sciences-Mongolian Paleontological Centre joint expedition (b). P. 7: L. Witmer (t);
J. Hutchinson (c); S. Hartman (b). P. 8: D. Hone / Royal Tyrrell Museum of Palaeontology.

British Library Cataloguing-in-Publication Data
A catalogue record for this book is available from the British Library.

Library of Congress Cataloguing-in-Publication data has been applied for.

ISBN (hardback) 978-1-4729-1125-4
ISBN (trade paperback) 978-1-4729-1126-1
ISBN (ebook) 978-1-4729-1127-8

2 4 6 8 10 9 7 5 3 1

Illustrations by Scott Hartman

Typeset in Bembo Std by Deanta Global Publishing Services, Chennai, India
Printed and bound in Great Britain by CPI Group (UK) Ltd,
Croydon CR0 4YY

*Bloomsbury Sigma, Book Thirteen*

# Contents

Preface                                              7

Note From The Illustrator                            9

The Game of The Name                                11

A Brief Primer on Tyrannosaur Bony Anatomy          17

PART 1: INTRODUCTION

Chapter 1: Introducing the Dinosaurs                 23

Chapter 2: What is a Tyrannosaur?                    35

Chapter 3: Tyrannosaur Species                       51

Chapter 4: Tyrannosaur Relationships                 63

Chapter 5: Tyrants in Time and Space                 73

PART 2: MORPHOLOGY

Chapter 6: Skull                                     87

Chapter 7: Body                                      99

Chapter 8: Limbs                                    111

Chapter 9: Outside                                  123

Chapter 10: Physiology                              133

Chapter 11: Changes                                 147

PART 3: ECOLOGY

Chapter 12: Reproduction and Growth                 161

Chapter 13: Prey                                    177

Chapter 14: Competitors     199

Chapter 15: Obtaining Food     219

Chapter 16: Behaviour and Ecology     237

PART 4: MOVING FORWARDS

Chapter 17: *Tyrannosaurus* Fact and Fiction     251

Chapter 18: The Future     261

Chapter 19: Conclusions     273

References     281

Further Reading     290

Museums and Online Sources     296

Acknowledgements     299

Index     300

# Preface

When writing a science book of close to 90,000 words long, you can be pretty sure of two things. One: it will contain at least a couple, and probably several, pretty bad errors; and two: it will be out of date by the time it is published. The first I am largely resigned to, the second is more important to understand. The rate of discovery and scientific exploration of dinosaurs is accelerating all the time to the point that, on average, a new species is named every week or so; this is in addition to all the other new studies and assessments of behaviour, anatomy, ecology, evolution and the like that will be published in that time.

Since the tyrannosaurs were first recognised as a group of dinosaurs in 1905 with the naming of the genus *Tyrannosaurus*, a huge number of scientific works have been written about them. A quick search of my own, far from comprehensive digital library finds more than 1,500 papers and books that relate to the subject. One famous paper is nearly 150 pages in length and is primarily a description of a *single* skeleton. We do know an awful lot about tyrannosaurs, and that knowledge is expanding faster than I have been able to write this book. Already I have had to add to or rewrite sections several times to take account of new ideas, new data and even entire new species that have been described while this book was being composed.

This is not intended as a text book or reference work, so I have skimmed over citing much of the formal scientific literature that forms the basis of our knowledge of the tyrannosaurs. An exhaustive list of papers on tyrannosaurs alone might be longer than this book, so I couldn't include them all, even if I had wanted to. However, it is important to try and mention key papers, and to show the scientific basis of the ideas and hypotheses laid out over the course of this work. Where appropriate, numbered references to sources are given in the text, and the sources are listed in full at the back of the book. While the reference list is effectively far from complete, as far as possible the information here is backed by at least some scientific studies (a number, of course, are controversial, uncertain or even contradictory), except where I have made it clear that something is based primarily on my own opinions and ideas. Even so, for every paper in the reference list there are perhaps dozens more that explore the ideas expounded upon, and similarly for every fossil illustrated or mentioned, there may be dozens or hundreds more specimens that have been the subject of study or analysis to support an idea.

In this book, I have attempted to steer a middle course covering primarily what I think is the consensus opinion among dinosaur researchers. While minority ideas get a look in, the scope of the book limits the discussion of some areas. I have tried to streamline and simplify often difficult and complex issues, but without overlooking nuance or important contradictions, and to give credit as far as possible given the imposed limitations on space – the intention is to provide a fair and balanced look at exactly what *I* think the tyrannosaurs were like.

# Note From The Illustrator

Scientific illustration of fossils is a curious thing. The goal is to provide the most accurate possible view of extinct organisms, in this case tyrant dinosaurs and their relatives. Naturally, this can't be a literal view – after all, some of the bones are usually missing, and many of those that survived were distorted by being squashed under tons of rock for tens of millions of years. To put the skeletons together in a literal fashion would render them incomplete and often oddly twisted upon themselves.

Instead, I have endeavoured to provide the most accurate possible portrayal of the skeletons as they were in the living organism. To this end, the first job was to get the proportions right. I have tried to take my own measurements of the original bones where I can, and where expense and distance have made that prohibitive, I have relied on measurements supplied by colleagues. They have all been checked against the vast and ever-growing scientific literature on dinosaurs.

Illustrating dinosaur bones in the proper shapes and proportions is all well and good, but a full skeletal reconstruction requires their arrangement in a manner consistent with the living animal. For this I have relied on detailed observation of specimens, the vast scientific literature on biomechanics and functional morphology, and dissection of extant organisms for comparative purposes.

Wherever possible, I have tried to make the creation of anatomical diagrams a data-driven process, though naturally there are limitations to our state of understanding. Missing data has been filled in from other specimens of the species or their close relatives, and logical anatomical inference was used when hard data was missing (Fig. 1). Future discoveries will undoubtedly require revisions of the interpretations.

## *Tyrannosaurus rex* comparison

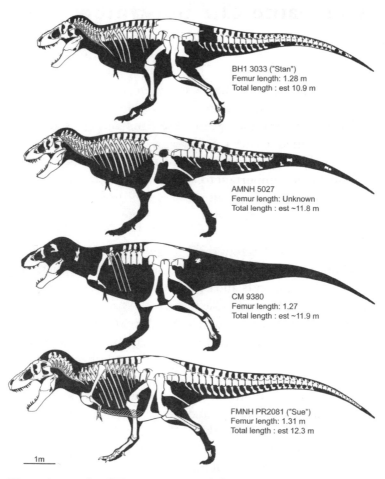

*Fig. 1 A panoply of* Tyrannosaurus *skeletons. Few dinosaur fossils are truly complete with every bone known and in good condition, but there is substantial overlap between many and so we can have great confidence in restoring the complete skeleton from these incomplete remains.*

Despite any yet-undiscovered errors, every attempt has been made to provide a visual representation of these extinct beasts that matches our current understanding of them. I hope you find them as fascinating as I do.

Scott Hartman

# The Game of The Name

Throughout this book there are references to various scientific names of tyrannosaurs, and indeed other dinosaurs – both individual genera and species and also formal evolutionary groups (termed 'clades'). Some of the terminology and rules for these names and their creation may seem complex and obtuse, but it is important to use them. Scientific terminology is there precisely to provide something that is specific and accurate, and not ambiguous. There's little point in reinventing the wheel.

Few of the old taxonomic ranks of organisms (Kingdom, Phylum, Class, Order, Family, Genus and Species) are used by modern biologists and palaeontologists. While terms like 'the dog family' and 'Class Aves' still get bandied about, researchers are increasingly abandoning them since they don't have clear equivalents between groups. We do still think of groups within groups (so all humans are apes, all apes are primates, all primates are mammals and so on), and the technical names are still used to designate those clades (*Homo*, Hominidae, Primata, Mammalia), but the ranks are not considered a part of this.

The exceptions are the genus and species names, the traditional scientific or binomial name (often called a Latin name). A few of these at least will be recognisable to many people, for example *Homo sapiens*, *Boa constrictor* and, yes, *Tyrannosaurus rex*. A species is the basic unit of biology, and effectively denotes a group of individuals that are more closely related to each other than to any other individuals (other species). Biologists actually have a hard time defining species due to the bewildering variety of organisms out there and the fact that in the course of evolution species and lineages are constantly changing. The individuals that make up the species *Boa constrictor* right now, are not those that will be around ten or a hundred years from now, or those that were around a thousand years ago. Species ultimately blur into one another, though of course that's generally hard to see on the scale of a human lifetime or in the fossil record.

When species are named they are assigned to a genus, and the two names are used together to correctly identify an organism, which is important as there may be multiple species within a single genus. In the case of dinosaurs almost all genera contain just a single species, and as a result researchers (and the public) tend to refer primarily to the generic name – hence *Triceratops* and *Diplodocus*, but not often *Triceratops horridus* or *Diplodocus carnegiei*, or for that matter *Diplodocus longus*. A reduced form of the name is often used, which only includes the initial of the genus – thus *Tyrannosaurus rex* would become *T. rex*. Genus and species names are both italicised, and the generic name but not the specific one gets a capital. In the case of the tyrannosaurs, every genus named except *Alioramus* currently has only a single species in it, so for simplicity only the genus name is generally used as it is unambiguous (for example *Tyrannosaurus* rather than *Tyrannosaurus rex*).

The definition of a species you might well recognise (a group of organisms capable of breeding with one another) certainly is *one* definition of a species in use, but it's not much use for asexual bacteria, or for that matter extinct animals that are known only from fossils, so plenty of other definitions are also in common use. In the case of fossils, palaeontologists use a 'morphological species concept' – in short, they ask whether the organisms have a series of consistent anatomical differences that are likely to have been reflected in the living species. So, for example, when identifying different fossils, size is not a good basis for comparison (lots of species have markedly differently sized individuals in their ranks), nor are subtle differences such as having one more or one fewer teeth. However, a major difference like having much longer legs, or three fingers instead of four, or a crest on the skull, can be more convincing. Even so, there can be disagreements about whether or not two skeletons are effectively 'different enough' to warrant being named as separate species, or separate genera, and the criteria tend to be different between groups, depending on the data available and to a degree the researchers doing the work.

## Who is related to whom

In the last 20 years or so, a great deal of work has gone into working out the relationships between species, including for dinosaurs. Essentially the various characters of an organism are tallied and compared with other species. Those with the most features in common are considered the nearest relatives of one another, since they have diverged least from their common ancestor. In the case of the tyrannosaurs, there are five main groups to deal with – the tyrannosauroids and tyrannosaurids, and then the protoceratosaurids, albertosaurines and tyranno-saurines (see Fig. 2). Some groups lie within others, so all tyrannosaurines and albertosaurines are also tyrannosaurids, and all tyrannosaurids are tyrannosauroids. We can also use these terms to split off the non-included genera, for example by talking about non-tyrannosaurine tyrannosaurids. This system appears a little unwieldy at first, but it is relatively easy to get used to.

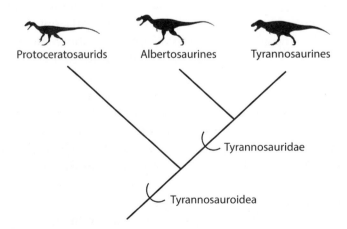

*Fig. 2 A very simple phylogeny of the major tyrannosaur clades.*

Throughout this book the terms 'tyrannosaur' and also 'tyrant dinosaur' are used to be essentially synonymous with 'tyranno-sauroid', encompassing all the animals shown in Fig. 2. These are not formal scientific terms, but they are convenient and I think appropriate. Palaeontologists use the term 'non-avian dinosaurs' because birds are really dinosaurs – here, however, to help readability, the name 'dinosaur' encompasses only the 'traditional' dinosaurs and excludes the birds.

## There's a time and place for everything

It is important to place the tyrannosaurs in the correct temporal and geographic contexts if we are to understand them fully. The world they occupied (one could even say 'worlds', given the times and changes involved) was very different from the one we would recognise today. The continents lay in different locations, the climate was different, the species around them (competitors, prey, plants, parasites and so on) were different, and these factors all influenced how the tyrannosaurs lived, evolved and died.

The surface of our planet is forever changing. Although a look out of your window may not reveal much difference from day to day, the twin effects of erosion and deposition are ever at work: taking material from one place and depositing it in another. Occasional dramatic events like floods, volcanic eruptions and rock slides can move thousands of tons of material in minutes, but mostly such things happen too slowly to be seen or appreciated on a human timescale – given enough time, however, the mountains really will crumble to the sea.

The movements of the continents are even slower. In the course of occasional 'exciting' events like earthquakes, continental plates may move a number of metres in seconds, although the rate is more normally a few millimetres over a year. Again, however, these events are hard to appreciate over a lifetime that is rarely as long as a century – yet with the dinosaurs we are often looking at continental changes that took place over millions or even tens of millions of years.

Importantly, factors like the positions of the continents influence local climate and weather, and also limit or allow movement between land masses for various species. It should also be borne in mind that due to the movements of the continents, many fossils may have moved from their locations in the past. In the Jurassic it would have been relatively easy to walk from South America to Australia via Antarctica, not only because the climate was much warmer then, but also because those land masses were joined together.

Some of the time spans are colossal and hard to imagine. The first dinosaurs arose around 240 million years ago (mya) in a period called the Late Triassic. The first tyrannosaurs did not come onto the scene for quite some time, with the earliest that we know of appearing in the Middle Jurassic period around 80 million years later (160 mya). The last of the non-avian dinosaurs, including the last of the tyrannosaurs, went extinct at the end of the Late Cretaceous period around 66 mya, so tyrannosaurs of one form or another were around for the best part of 100 million years.

Starting in the Late Triassic through to the End Cretaceous, the continents have shifted from something close to a single great land mass, to a layout not too far from the present one. The climate has (overall) cooled since then – the Triassic lacked ice at the poles – and the life that occupied these lands has changed dramatically, too. A visitor to the Late Triassic would recognise a number of plants and animals as being close to those still alive today. Early mammals that were rather rat- or possum-like would be scurrying around, there would be lizards, tortoises and crocodiles, and plenty of recognisable insects, spiders, millipedes and other invertebrates, while ferns, cycads and horsetails would have been common. Non-dinosaurian ancient reptiles such as pterosaurs in the air and icthyosaurs in the sea would also abound from the Triassic onwards, while in the Jurassic and Cretaceous the flora and fauna would become still more familiar, with lineages such as birds, snakes, grasses, magnolias and monkey puzzle trees appearing and becoming common.

As we will see, the place of the tyrannosaurs within these ecosystems changed dramatically over the 100 million years, from small and probably rare members of communities, to by far the largest carnivores on the continents they occupied, with the capacity to tackle almost any species around. The changing environment, and the changes to other species around the tyrannosaurs, would have influenced their evolution (as the tyrannosaurs in turn would have influenced those around them). The separation in time and place between some animals (the time difference between the respective appearances of *Guanlong* and *Tyrannosaurus*, is greater by a fair margin than the time between the last of the *Tyrannosaurus* and you reading these words) makes it important to recognise the sheer scale of the ancient past and how this may have shaped long-dead lineages.

# A Brief Primer on Tyrannosaur Bony Anatomy

I have limited the amount of anatomical detail and technical terms in this book. People often dismiss technical terms and names as 'jargon', but the truth is that a good technical term can be exact and specific, and avoid confusion and long-winded descriptions. 'Arctometatarsus' is a tongue twister of a word when you first encounter it, but it's a lot easier to refer to than 'that special condition when the middle long bone in the foot is pinched at the upper end and flares out at the base' every few lines when discussing its evolution and anatomy. There are therefore some anatomical terms in the text, and an introduction to them follows. Although every part of the skeleton has a name and is often associated with still more terminology relevant to its size, shape, orientation and which other parts it links to, only the pieces that most often come up in the discussion of the tyrannosaurs are included here. Names in bold below are marked on the illustrations.

Here the head of the legendary *Tyrannosaurus* is pictured conveniently side on, or more formally, in left lateral view (Fig. 3a). The skull as a term covers all the bones that make up the head, but can be split into the fundamental units of the **mandible** (lower jaws) and **cranium** (everything else). Key bones in the skull are those that hold the teeth: at the front of the cranium is the **premaxilla**, and behind that the **maxilla** (one on each side), while the front of each side of the mandible is a **dentary**. The nasals lie behind and between the maxillae and help bound the **nares** (the openings for the nostrils in the skull). Behind the naris on each side is an **antorbital fenestra**, then the **orbit** (the eye socket) itself. The roof of the mouth (not shown) is termed the palate. Each tooth has a root that sits in the jaw, and a crown that is exposed. The 'edge' of the tooth is a carina made up of tiny serrations called denticles.

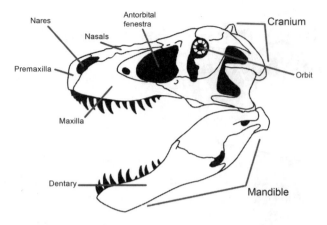

*Fig. 3a The major bones and features of the skull of a tyrannosaur, here based on* Tyrannosaurus.

Moving on to a whole *Tyrannosaurus* (again in left lateral view), it can be seen that tyrannosaurs fit with anatomical orthodoxy, with the skull being located at the front of the animal (Fig. 3b). Moving on to the axial skeleton (the backbone and associated bits), there is the **atlas**, that is the first vertebra of the neck (though it's hidden in the figure behind the back of the skull). Behind the atlas are the rest of the **cervical vertebrae** (neck) and their associated cervical ribs. Then comes the actual back part of the backbone, the **dorsal vertebrae** and their **dorsal ribs**. Lying along the belly is a series of fine bony rods occasionally called belly ribs, but better named **gastralia**. Posterior to the dorsals are the **sacral vertebrae**, which are fused together to compose the sacrum and make up the major part of the pelvis. After this comes the tail, made up of **caudal vertebrae**, and below them the **chevrons** (also called haemal arches in mammals).

Finally there is the appendicular skeleton, that is the limbs and their supporting girdles. Upfront are the shoulder elements of the **scapula** and **coracoid**, and between them the **furcula** and **sternum**, then the **humerus** (the upper arm bone), then the lower elements of the **radius** and **ulna**, and the wrist (**carpals**), the hand (**metacarpals**), and the bones of the fingers (**phalanges**), terminating with the **unguals** (the bony

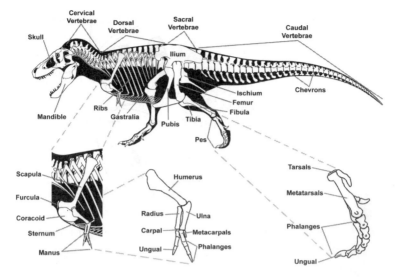

*Fig. 3b The major bones of the skeleton of a tyrannosaur, here based on* Tyrannosaurus.

claws). The pelvis supports the hindlimb and consists of three pieces on each side: the **ilium** at the top (hiding the sacrum to which it is fused), the **ischium** to the rear and the **pubis** to the front. The legs each consist of a **femur** first, then a large **tibia** and smaller **fibula**, then the **tarsals** (ankle bones), then the **metatarsals**, followed again by **phalanges** and **unguals**.

That is the skeleton in brief. In all there are a couple of hundred bones in a single individual tyrannosaur (and the best part of 100 teeth sticking out of the jaws of most) with the majority occurring in pairs (left and right arms, left and right ilia, left and right maxillae and so on), except the vertebrae and the furcula (though this is actually a pair of bones fused together: the clavicles or collarbones). Most fossil specimens are only a small fraction of this total count, and even skeletons described as being 'complete' may have a good number of bones missing (though of course if you have a left arm, you know what the right one looked like and so on).

There are some features unique to tyrannosaurs compared with other dinosaurs (like the hugely expanded free end of the pubis – termed the 'boot'). However, the overall plan of the skeleton is typically dinosaurian, and is pretty typical of most tetrapods (the group that includes all amphibians, reptiles, birds and mammals). Those familiar with human anatomy or that of other animals, like mammals and birds, will recognise a lot of these names as a result. The names are mostly shared because in an evolutionary sense, the bones are the same (although because anatomical terms were developed separately by various people at various times, mammal, bird and reptile anatomists can use different terms for the same bones). The skull has most of the same bones in the same places, the backbone is the same, there are four limbs, each with one bone, then a pair, then some smaller ones to form a joint, then fingers or toes. The evolutionary pattern of the tetrapod skeleton was set early on and remains pretty similar in almost all animals, with few gains along the way (for example, many birds have long necks with numerous extra vertebrae), and a fair number of losses (we barely have a tail, and snakes and whales have all but lost their legs). What dictates the difference between various animals is the shape and structure of these elements, and how they link to muscles, blood vessels, lungs and other anatomical features to build different-looking (and acting) species.

# PART ONE

# INTRODUCTION

# CHAPTER ONE

# Introducing the Dinosaurs

A round 160 million years or so ago, probably somewhere in the northern hemisphere, a new lineage of carnivorous dinosaurs began to separate from its relatives. They were probably not that different from their contemporaries to look at – after all, evolution doesn't happen overnight. These animals were pretty small, and didn't have especially big heads, long arms or short tails. They were just carnivorous bipedal (two-legged) dinosaurs out looking for a meal, taking what meat they could find and staying out of the way of the bigger beasts that populated the landscape.

However, given time, this new line of animals would come to include the largest terrestrial carnivores of all time, with giant heads, huge teeth, tiny arms and highly modified feet. The biggest individuals were perhaps around 14 metres in length and weighed over 8 tonne. They would be the subject of more research than any other lineage of dinosaurs (aside from the birds), and they would become icons of evolution, and of the dinosaurs as a whole. These were the tyrannosaurs – *Tyrannosaurus* and its ancestors and relatives – a group of dinosaurs that lived around the world for some 100 million years, represented by at least 25 known species. Perhaps no scientific name is as well known as *Tyrannosaurus rex* (and I am certain none is incorrectly written so often), and the animal that bears that name is easily the most popular and best-known dinosaur with the general public. *Jurassic Park* and *King Kong* would not have been the same without it, and visitor numbers increase every time a museum displays a new skeleton or animatronic model. It is hard to put a price on fame, but a skeleton of the largest and most complete individual *T. rex* known sold for millions of dollars in 1997 (Fig. 4).

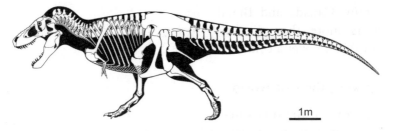

*Fig. 4 A restored skeleton of* Tyrannosaurus rex *based primarily on the famous and large specimen known as 'Sue'.*

The tyrannosaurs are endlessly charismatic: their size, appearance and reputation precede them wherever they go, and there is almost no media article or report on dinosaurs that doesn't mention them or relate any new find to tyrannosaurs in some way ('as large as', 'a relative of', 'lived 75 million years before', and on and on). There is also a huge number of tropes and memes about the animals that follow them around, and endless repetition does not make them right. Tyrannosaurs were not pure scavengers; they didn't spend their lives battling adult *Triceratops*, they did not have poor eyesight, they could not run at 50 km/h, females were not bigger than males, and so on. Unfortunately, the hype and hyperbole overshadows a fascinating evolutionary story: how and why did they get so large, why the tiny arms, why the giant heads, how did they act? We know so much about the animals in this group – their anatomy, evolution, behaviour and general biology – but it's almost impossible to say very much over the chorus of statements about how cool they are or questions as to whether they would win in a fight with *Spinosaurus*.

Exploring the wealth of scientific data and analyses will take in bones, skin, feathers, footprints, scans of tyrannosaur skulls, possible evidence of cartilage surviving for 66 million years, their muscle attachments, their injuries and their bite power, and even studies of entire ecosystems. There are skeletons and teeth from Australia through Mongolia, from

Britain, Canada and Brazil; the story of the tyrannosaurs covers more than 100 million years of palaeontological history, and over a century of scientific research.

## Awaking the lost world

The story of the dinosaurs begins in Britain in the early to middle part of the nineteenth century. In 1824, the then Reverend (and later Canon) William Buckland, a British naturalist based at the University of Oxford, named the first ever dinosaur: *Megalosaurus*. Known for some time from little more than a lower jaw and some teeth, it was still an incredible find, and a sensation at the time. The teeth showed quite clearly that the animal was a carnivore, and the fact that it came from what was called the 'age of reptiles' and had at least some features in common with crocodiles, pointed to a reptile of some previously unknown kind. Early reconstructions showed a giant, rhinoceros-shaped lizard with sharp teeth. People had already been finding great sea monsters (what we now know as ichthyosaurs and plesiosaurs), especially on the south coast of England, but now there was a giant land reptile too.

Not until more remains of *Megalosaurus* and the bones and teeth of other terrestrial saurians were unearthed in England would the term 'dinosaur' actually be coined. Sir Richard Owen (founder of the Natural History Museum in London, and brilliant anatomist and biologist, but not the nicest of people) realised that a number of these new animals had certain characteristics in common and could be grouped together, and were therefore worthy of a new name. Thus the name 'Dinosauria' ('terrible lizards') was coined in 1842, and the public has been entranced with them ever since. Science and natural history were popular pastimes in the nineteenth century, and new discoveries could become overnight sensations. Public lectures on palaeontology and ancient animals were standing room only, and exhibitions in Europe and the US had people coming in their thousands to see reconstructions of dinosaurs. They were popular from the word go, and rapidly became something of a fad for many in society. Although some of the other first dinosaurs to be discovered (notably *Iguanodon*, described by Owen's great rival, the doctor Gideon Mantell) were known from more complete remains and more numerous specimens, *Megalosaurus* was clearly first among equals in the public imagination. It is perhaps inevitable that the carnivores, with their claws and teeth, struck deeply into the consciousness. *Megalosaurus* is mentioned in Dickens's *Bleak House*, and is the key dinosaurian antagonist that threatens the explorers of Conan Doyle's *Lost World*.

Quite what message the public often got about these animals varied, however, as from the earliest discoveries there had been strong disagreement about the general appearance and lifestyle of the dinosaurs. Were they merely lizard-like reptiles writ (very, very) large, or something rather different and special? They were certainly huge, but perhaps also more like mammals and birds in some respects. The shapes of their hips and femora suggested that they had an upright stance rather than sprawling like the 'lower' reptiles, such as modern lizards (with the legs held out to the sides, not under the body), and it was quickly clear that *Megalosaurus* and its carnivorous relatives – the group would later become known as the theropods – walked on just their back legs. Although in Europe the prevailing idea seemed to be of relatively slow and sluggish dinosaurs, as championed by Owen, new discoveries and ideas in North America would soon shift the focus to the New World.

Plenty of dinosaurs have been discovered in Britain over the years, including many famous and important fossils, and it remains a major centre of dinosaur research. However, it soon became obvious in the mid-nineteenth century that a great many new and wondrous animals were coming out of the Americas. By the 1850s, new dinosaurs were being named from the United States, and the great excavations of the 1870s and '80s from the Midwest became as much of a story as the animals they uncovered. The 'Bone Wars' between palaeontologists Edward Cope and Charles Marsh produced dozens of skeletons, and many subsequently famous names (such as *Diplodocus, Stegosaurus, Allosaurus* and *Apatosaurus*) stemmed from their work. Each new expedition by one or the other seemed to produce a larger and more incredible animal than the last. Although these rivals were the key protagonists, other important researchers were active in the late nineteenth century and into the early part of the twentieth. George Sternberg and Barnum Brown are among the most notable of these collectors, and the latter in particular is one to whom we must pay special attention – he was the man who found the remains of what was soon to be named as the first tyrannosaur – *Tyrannosaurus rex*.

However, the story of tyrannosaurs actually pre-dates Brown's biggest moment in an already stellar career. Joseph Leidy named what was arguably the most complete specimen known at the time, christening the first hadrosaurian (duck-billed) dinosaur *Hadrosaurus* back in 1858 (unlike other early finds such as *Megalosaurus* and *Iguanodon*, this included most of a skeleton). Like both Owen and Mantell, Leidy was a medical man by training who moved into anatomy and also made contributions to palaeontology and mineralogy. Having an interest in fossils, he was able to obtain some of the first bones and teeth to have been excavated in the New World, and went on to publish a series of papers on them. In 1855, explorer Ferdinand Hayden collected several teeth in the Judith River region of Montana, an area we now know to be extremely rich in vertebrate fossils from the Late Cretaceous, and had them shipped to Leidy.

The following year, Leidy gave descriptions and names to a number of these teeth. At least one of his names is still considered a valid one: the small carnivorous dinosaur *Tröodon* (though now known without the diacritic mark) – but a number of others have passed into history. Further discoveries have shown that, while clearly novel at the time, none of the features on them are consistent enough or clear enough to accurately define a species. Among these now lost names was *Deinodon*, meaning 'terrible tooth', which Leidy suggested was a close relative of *Megalosaurus*. It's easy to see why: although fragmentary, the teeth are obviously from a carnivore, being curved and having a serrated cutting edge, and with few other dinosaurs known at the time it was a reasonable comparison to make. However, what stands the teeth apart is the colossal size of those of '*Deinodon*', being more than twice as big as those known at the time for *Megalosaurus*. We are now relatively sure that the '*Deinodon*' teeth belonged to a tyrannosaur, based on the age of the specimens and their place of origin. However, they are anatomically indistinguishable from the teeth of several tyrannosaur species – which is why the name *Deinodon* is no longer considered valid, and also why we are not sure exactly what the teeth belonged to.

What are probably the first remains of *Tyrannosaurus* itself to be recovered now reside at Yale, and are represented again by a large tooth. This one was found by teacher Arthur Lakes near Denver, Colorado, and sent to Marsh in 1874. Although the Lakes tooth was not named, a number of other fragmentary specimens were christened but are now referable to *Tyrannosaurus*. These include *Manospondylus*, named by Cope, and *Dynamosaurus*, which was named in 1905 by yet another great American palaeontologist of the 1800s, Henry Fairfield Osborn, alongside *Tyrannosaurus* itself.[1] The rules of naming are such that in cases where the same genus or species has been named twice (or more), the first name takes priority. Had Osborn named his species in a different order, this book might have been called *The Dynamosaur Chronicles*.

Although *Dynamosaurus* and *Deinodon* have not survived, Osborn had plenty of other successes, and the tyrannosaurs were on their way. Osborn named many new dinosaurs that are still valid, including the tyrannosaur *Albertosaurus*, which unsurprisingly comes from Alberta in Canada. This animal was found by Joseph Tyrrell in 1884 when prospecting around the Red Deer River, and stimulated a new rush of dinosaur excavations in what turned out to be a tremendous region for Cretaceous fossils, and in particular for tyrannosaurs.

The area is now a World Heritage Site named Dinosaur Provincial Park, and it is also home to the Royal Tyrrell Museum of Palaeontology, whose name honours one of Canada's greatest collectors. Because of the erosive action of the river, several expeditions were actually conducted by boat, moving up and down the waterways to access new areas and haul dinosaurs from the cliffs. One of these trips turned up a quarry containing numerous tyrannosaurs, but this was not excavated until detective work led to its rediscovery in the late 1990s, and the unearthing of more than 20 *Albertosaurus* skeletons.

*Albertosaurus* was the first of several relatively large tyrannosaurs to be described from Alberta. Excavations by Barnum Brown for the American Museum of Natural History and the Sternberg family led to a huge new series of dinosaur discoveries from the region in the early 1900s. In 1906, Brown unearthed a new *Tyrannosaurus* specimen that included a near-complete skull and much of the rest of the animal (a specimen now on display in the American Museum of Natural History in New York), and gave the first great view of the tyrannosaurs beyond some teeth and jaws. Also from Alberta came *Gorgosaurus*, entering into the scientific registry in 1914 and, rather later, *Daspletosaurus* in 1970. However, overall tyrannosaurs remained rather elusive for much of the twentieth century, and for a long time these discoveries represented much of what we knew in terms of the diversity of the group.

## Twentieth-century carnivores

The flurry of dinosaur research that had typified the late nineteenth and early twentieth centuries slowed, if not to a crawl, then certainly to a much lower rate for the next few decades. There was less fieldwork, and fewer dramatic new finds. Fields like biology accelerated as our understanding of genetics and inheritance developed, and behaviour began to be studied in a systematic manner, while new technologies helped drive new experiments in physics and chemistry.

In palaeontology, however, the traditional cores of taxonomy, anatomy and fieldwork, based on people with shovels, continued. In addition to the interruptions of the world wars, more obvious fossil localities in Europe had been exhausted; those in North America were still productive, but mostly producing more examples of already known species. Africa was for various reasons a difficult place to work in, South America was still largely unknown, Australia seemed all but bereft of dinosaur fossils (a notion only recently found to be grossly incorrect), and the major parts of Asia known to be good for dinosaurs (eastern Russia, Mongolia and China) were locked to the West by political divisions.

However, some palaeontological work did take place in the Far East. Starting with the famous Central Asiatic Expeditions in the 1920s, Roy Chapman Andrews (a man often incorrectly, though understandably, thought to be the model for Indiana Jones) and his team found the first dinosaur eggs and many brilliantly preserved dinosaurs in Mongolia. Whole new animal groups were described from these new discoveries (including the now notorious *Velociraptor*), and not just dinosaurs, but important early mammals as well. However, the remote and difficult location of the fossil sites, as well at the turbulent political situation, limited the work. With first the Japanese invasion of China, then the Second World War and the Cold War, visits by Western scientists stopped, even if those by Eastern Bloc researchers could continue. Russian, and in particular Polish, researchers working in Mongolia were unearthing large numbers of new dinosaurs, among them many tyrannosaurs. First came the giant *Tarbosaurus*, formally named in 1955 by Russian palaeontologist Evgeny Maleev, and later the first material of the unusually slender *Alioramus*. By the 1980s restrictions were being eased, and Western researchers began to again explore these areas and collaborate with their counterparts in Asia. There had been major advances in various scientific fields in the West that had not always trickled eastwards, and while Asian research had been productive in terms of finding and naming new animals, it was not always in the context of ideas that had progressed in North America and Europe.

Starting in the 1970s and moving through to the 1980s came the shift that has become known as the 'dinosaur renaissance'. Although research had been progressing along various lines since the first dinosaurian discoveries of the mid-nineteenth century, for the majority of the time since the dinosaurs had been considered to have been largely reptilian in nature. The old images of them lounging neck-deep in foetid tropical swamps, or waddling around the landscape with their tails dragging along the ground, are a product of this mindset. Dinosaurs were regarded more or less as giant lizards, and they were supposed to be animals of very little brain, scaly and incapable of operating unless the temperatures were high. Sure, they achieved huge sizes and dominated the terrestrial realm for a considerable period of time, but they were still just big reptiles.

## The dinosaur renaissance

At this time, more and more work was pointing towards two key, and interlinked, ideas. The first was that the dinosaurs were rather more active than we previously realised. Footprint tracks did not show drag marks from tails, and some footprints were so far apart the animals that made them (even the larger ones) were clearly running. Fossils showed that various dinosaurs inhabited polar regions, which were cold when the dinosaurs were alive – they were not restricted to tropical climes like most modern reptiles. The large dinosaurs grew very fast, reaching adult sizes quickly. Indeed, their very anatomy enabled them to be upright walkers, rather than lizard- or croc-like sprawlers. These factors all pointed to active animals with elevated metabolisms that could be active and fast moving at any time of day and in any temperature, by no means limited solely to the heat of the day. This general concept had been floated in the early days of dinosaur research (in part by Gideon Mantell, though

Richard Owen's views generally held sway), but despite some good evidence to support it, the idea had quickly fallen by the wayside.

Still more critical was the second key concept, the idea that birds may have descended from dinosaurs. Oddly enough this had also made an early appearance, with some work in the early 1900s pointing to the similarities between the famous feathered *Archaeopteryx* from southern Germany and some of the carnivorous dinosaurs. Now, however, the idea came to the fore again, led in part by the discovery of new theropods that prompted renewed comparisons to birds, with an increasing number of avian features being found in non-avian dinosaurs.

Now the dinosaurs as a whole were getting a really new shot in the arm (or perhaps wing). If dinosaurs were the ancestors of birds, and non-avian dinosaurs had a number of bird features, then some of the characteristics we traditionally ascribe to birds could, even should, be present in the Mesozoic. Dinosaurs didn't have to be just reptile-like; they could have been bird-like in some, or indeed many, regards. They may have been more intelligent and more active, and have engaged in extended parental care of their young, as well as other activities and behaviours.

Research soon accumulated in both areas, and by the end of the 1980s the majority of palaeontologists had come round to the idea. Birds are real-life, living dinosaurs, and the dinosaurs were not lumbering lizards as had initially been imagined, but active animals like the birds of today. Since then huge strides have been made in research, and many avian features have been revealed in various dinosaur lines (though of course mostly in the theropods, from which the birds are descended). We now have evidence of a plethora of feathered dinosaurs, and of brooding on nests, air-filled skeletons and numerous anatomical features that dinosaurs directly share with birds.

This was the beginning of a new wave of interest in dinosaurs and the Mesozoic in general, with research levels increasing rapidly. Renewed interest in previously underexplored areas such as China, Argentina and Australia started to generate huge amounts of new fossils, even previously productive areas threw up new finds – and all of this was combined with the identification of interesting things in old museum collections that had been overlooked or not understood before. The number of researchers working on dinosaurs increased, and the rate of research bloomed. In particular, the rate at which new species were identified went through the roof.

Naturally, the tyrannosaurs benefited from this as much as other lineages (and quite probably more, as they have always held a certain cachet). As a result, new batches of tyrannosaurs began to appear. An early tyrannosaur, *Eotyrannus*, popped up in Britain in 2001; the first feathered tyrannosaur, *Dilong*, was named in 2004; *Proceratosaurus* (named in 1926) was identified as a tyrannosaur in 2010; and discoveries are still being made, with *Zhuchengtyrannus* described in 2011, *Yutyrannus* in 2012, *Lythronax* in 2013, and *Nanuqsaurus* and *Qianzhousaurus* in 2014. Public interest in dinosaurs in general, and in tyrannosaurs in particular, continues apace. When a near-complete skeleton of an especially large specimen of *Tyrannosaurus* came to light in the 1990s (the now near legendary 'Sue')[*], it made headlines around the world, and when the ownership of the specimen was disputed and the material seized before being auctioned for a colossal eight million dollars, the media ran with the story almost endlessly.

We are, then, in a real golden age of dinosaur research – and therefore also of the tyrannosaurs. We now know more about their anatomy, evolution, ecology and behaviour than we did even ten years ago, and research continues apace to develop new ideas, confirm existing ones and reject some others. *Tyrannosaurus* was neither a lumbering scavenger, living off the dead and dying and crunching its way through the remainders of carcasses, nor some super-predator capable of sprinting after fleet-footed prey. But how did we get to *Tyrannosaurus*, that most iconic of organisms? What were the origins of this giant animal? Tyrannosaurs lived on Earth for

---

[*]Note that this is merely a nickname and does not relate to the possible sex of the actual animal.

100 million years, and produced dozens of species across several continents. What were the evolutionary patterns that took this once small and unremarkable set of animals and morphed them into one of the largest and most unusual predators of all time? These questions and others are tackled in the sections that follow.

CHAPTER TWO

# What is a Tyrannosaur?

Taxonomy and systematics (defining species and clades and working out their relationships) form the basis of our modern understanding of biology. Working out which species are which and how they relate to one another allows us to reconstruct the evolutionary history of organisms, and in the case of fossils, make inferences about them based on their living relatives or more complete remains of other extinct animals. If you cannot work out to which group an individual organism belongs it is difficult to say much about it in terms of its evolution, biology or how it relates to other individuals or groups. Thus it is important to work out what species are, and are not, tyrannosaurs, but also how the tyrannosaurs relate to the rest of the Dinosauria.

## What are the dinosaurs?

Defining what is and what is not a dinosaur is rather more complex than it once was – but it's also rather more rigorous. Before the advent of cladistics (the method by which we reconstruct the relationships of species and clades)*, groups were created by taxonomists based on a few key features that they felt unified them. Following the publication of Charles Darwin's *On the Origin of Species* in 1859, such features could be used to then infer a shared evolutionary ancestry. However, the problem here was that potentially the characters chosen might have evolved independently, and do not indicate a recent shared history. Groups like the 'pachyderms' were created, and while the term is now used as a synonym for elephants, this was originally considered a real taxonomic group consisting of the large, 'thick-skinned' (the literal translation of pachyderm)

*This is covered in more detail in the next chapter.

herbivorous animals – rhinos, hippos and elephants. We now know that rhinoceroses are closer to horses than these other animals, hippos are closer to deer and antelope (and whales as it happens) than to the others, and elephants are nearer to the odd little hyraxes, and the dugongs and manatees than to the rest. In short, this one major feature, although shared by these animals, isn't a key to their relationships.

By using a great many characters and searching for the species or groups with the most in common, cladistics avoids the trap of accidentally picking a couple of key features on which to base your taxonomy. Instead, we can build the hypothesised relationships first – the phylogeny – then see how the characters fall out, and select those that help define the groups. This also has the advantage of making sure you don't miss out things that should belong in a correctly defined group because the character you chose doesn't turn up in everything (whales are still mammals even if they have lost their hair). Even so, it is worth noting how much the taxonomists got right in the pre-cladistic area: many of the relationships throughout biology that have been strongly supported by cladistic work do generally match what had been suggested before, and indeed many of the characters that support those groups are still valid today. The modern methods are clearly more rigorous, but it's not like the old work was terrible.

Cladistics might make the identification of groups and characters associated with them seem rather redundant, but this isn't the case. It is very useful indeed to know what characters identify given groups (or species), as this allows us to then work on things (be it an odd bone in a collection, a new find in the field or just generally discussing details) without having to run a new analysis every single time to check. As a result, lists of characters that define groups are still commonplace and extremely helpful.

## Scaly skins and 'cold blood'

Let us start with the obvious question: just what is a reptile? Here we hit an immediate problem that is common when dealing with scientific terminology that does not immediately or obviously correlate with the popular idea of a group. Ask the average person on the street and they'll probably suggest that reptiles have scaly skin and are 'cold blooded', and perhaps add that they lay hard-shelled eggs. Those with a bit more knowledge might name lizards, snakes, tortoises and crocodiles as being reptiles, and bonus points would definitely be available to anyone who mentioned tuatara (if you don't know what these are do look them up, because they are fascinating animals). Such a description and list of animals is far from a bad start, though plenty of erroneous things like newts, or a forked tongue being ubiquitous, would also no doubt be suggested. However, there is a scientific definition of the clade Reptilia (sometimes now called Sauropsida), which reads rather differently from this and has rather different implications.

Reptilia is essentially considered to embrace all animals that lie on the branch of life closer to the living groups including lizards and birds, than the branch that produced the mammals. These animals do have features in common that are found (at least ancestrally) in the group, most notably scaly skin, although other characteristics like egg laying are also present in other groups ancestrally (don't forget that modern monotremes are mammals that lay eggs), so do not define the group. Also inherent in the definition of any evolutionary lineage is that all descendants of that lineage are part of that group. Dinosaurs are very much a member of clade Reptilia, but birds are descended from dinosaurs. Birds are therefore quite literally members of Reptilia, and by extension are reptiles. You can immediately see, then, where the disjunct between the common, anglicised form of the name 'reptile' jars with the scientific, technical one (hence in part the increasing use of Sauropsida as a name). Reptiles are a particular problem *because* the common name predates the scientific one, so for me at least it's hard to argue too much with the public's misunderstanding of the terminology and the difference between the two – this is thus one area where I find it useful to tread carefully to avoid confusion. The use of Dinosauria versus dinosaur, however, is quite different, since the technical name predates the common form, so I think it should have something akin to priority.

Returning to the main theme, Reptilia does contain a great many living and even more extinct lineages of animals that have roamed the land, sea and skies for hundreds of millions of years. The dinosaurs might be the most famous, and lizards and snakes the most familiar, but there were many weird and wonderful animals that lived before the dinosaurs and alongside them. Various large extinct marine animals such as plesiosaurs and icthyosaurs are most often mistaken for dinosaurs simply because they are large and extinct reptiles that lived in the Mesozoic, but there were also the tank-like, armoured aetosaurs, flattened-turtle-like placodonts, and beak-faced rhynchosaurs among others (particular favourites of mine are the truly incredible Triassic beasts *Tanystropheus* and *Drepanosaurus*).

## Ruling the roost

Within the reptiles is a key group – the archosaurs. The translation of this name, 'ruling reptiles', is a perfectly reasonable moniker for both past and present animals in this group. The archosaurs consist of crocodilians and their relatives, the flying pterosaurs, the dinosaurs and, by extension, the birds. The dinosaurs include the largest animals ever to walk the Earth and were the dominant clade for the best part of 150 millions years, while the birds comprise the most diverse group of tetrapods around today (around 10,000 species). The archosaurs definitely rule. The group is defined by the shape and structure of the skull and a few other key features, such as the presence of serrated teeth (obviously absent in modern birds) and the shape of the hip joint.

When it comes to the dinosaurs themselves, a definition is rather unexciting in a sense. A good fossil record for a group will provide a lot of information on intermediates: transitional or in-between forms and species within the group. One group doesn't suddenly stop and another abruptly come into being: evolution takes time to sculpt, shape and change things. A patchy fossil record can thus make one group appear to have sprung out of nowhere (as used to be the case for birds and whales, for example), but a superb fossil record (like the one that now exists for both of those groups, and also for the arrival of the dinosaurs), results in a 'blurry' and far from clear-cut transition. This is exactly what you would expect as evolution slowly modifies species and lineages from one form to another, but it also makes it hard to draw firm lines between groups, hence in part the complexity of defining species and clades.

Clade Dinosauria is rather special. Separating from the other archosaur lines around 240 mya, dinosaurs came to dominate the terrestrial realm of the Mesozoic. We have now named in excess of 1,500 non-avian species for this group (which is admittedly small when you consider that there are around 6,000 species of mammal alive now, in one tiny period of geological time) in about 35 major groups (previously called families). Dinosaurs are known from every modern continent (including Antarctica), and have been recovered from a huge number of countries, even those with little rock of the right kind, and those that might not be famous for their dinosaurs, such as Japan, New Zealand, Peru, Italy, Niger and South Korea. Mesozoic dinosaurs are found everywhere from Tibet to Tierra del Fuego via Belgian coal mines, Australian deserts and the Canadian High Arctic. They are a truly global group, and ranged in adult size from animals that could sit comfortably in the palm of your hand, to monsters that weighed tens of tons. They include bipeds and quadrupeds, carnivores and herbivores, species with horns, spikes, armour, huge crests on their heads and hundreds of teeth, and those that were fish eaters, ant eaters, filter feeders, burrowers, runners, climbers and even gliders. To write them off as giant lizards, which many people still do, is to grossly underestimate their diversity in form as well as their evolution and behaviour.

Looking at the dinosaurs in a little more detail, they can be split primarily into two clades, the ornithischians and the saurischians, with the major (but not the only) defining characteristic of each, the shape of the pelvis, reflected in their names. The ornithischians have 'bird-like' hips, with a major bone of the pelvis, the pubis, pointing backwards, while the saurischians have a 'lizard-like' hip with a forwards-projecting pelvis. Interestingly or confusingly, depending on your point of view, the birds are actually saurischian dinosaurs – in the later members of the saurischian groups the pubes were reversed before the evolution of birds, and this characteristic independently converged on the same pattern. Of course, these names were designated when no one had the slightest inkling that birds might have been related to the dinosaurs, and it is a good example of how things can change.

With the possible exception of some early forms, the ornithischian dinosaurs were all primarily herbivores, and some featured amazing adaptations towards the efficient acquisition and destruction of plant matter. They generally had both beaks at the front of the snout and teeth behind, giving different food-processing options in one mouth. The ornithischians include a huge diversity of forms, with some being armoured, plate-covered and spiky quadrupeds, some having massive horns and frills, and others being small and fast; there were bipeds and quadrupeds, and many species sported some of the largest and most elaborate crests on their heads for sexual and social displays. Dinosaurs with familiar names like *Stegosaurus*, *Triceratops*, *Ankylosaurs*, *Iguanodon* and *Hadrosaurus* are all ornithischians.

The saurischians are perhaps most diverse overall in terms of numbers, and certainly in terms of extremes, but are rather more conservative in their overall forms, with only two major body plans. There are two groups within the Saurischia and each is instantly recognisable in its own way. The sauropodomorphs are known primarily for producing the real giants, with a great many animals weighing more than 10 tons, and some of the largest very probably exceeding 50 tonnes (perhaps by quite a margin, and while some old estimates of 100 tonnes or more are clearly incorrect, some modern estimates still come in at close to 80 tonnes for the biggest sauropodomorphs). Early forms were actually rather like the earliest theropods, and indeed some species have moved around the base of the dinosaur tree between the sauropodomorphs, theropods and ancestral dinosaurs in various analyses. Some of these were probably predators or omnivores: opportunists taking whatever they could. However, this soon gave way to herbivory dominating, before things took off figuratively and almost literally in terms of reaching for the skies.

The early sauropodomorphs were still bipeds, but they soon evolved into larger and larger sizes, and shifted from walking on two limbs to four. Their necks mostly increased in length, so we are left with the classic forms like *Diplodocus*, *Brachiosaurus* and *Apatosaurus*: multi-tonne giants with huge bodies, columnar limbs and long necks supporting small heads, and often long tails. The sauropods reached their peak of diversity in the Jurassic, but in the Cretaceous were rather more reduced in numbers, but not in size, and were concentrated in the southern continents – though a few (including some real giants) survived in the north.

### Beast foot forwards

So to the second group of saurischians, the theropods: home of the tyrannosaurs. The name theropod means 'beast foot', and this should be a bit of a clue to what the theropods were getting up to. The ancestral dinosaurs were actually rather theropod-like in appearance, so we can perhaps consider the sauropodomorphs and ornithischians as being more divergent from the original stock than the typical theropod. All theropods were bipedal, and although resting prints show that they would take weight on their arms and pubis when sitting down,[2] they were fundamentally two-legged animals. The arms varied enormously in length, with some of the last non-avian theropods having extremely long arms relative to the rest of them, while in the tyrannosaurs and several other groups the arms were reduced to little more than nubs. Early theropods had four fingers on the hand, though this reduced to three in most, while a few had only two or even just one functioning finger, and in almost all cases they bore curved and rather sharp claws. Although theropods typically had four toes in the foot, the first was generally small and positioned high up on the foot, rather like the dewclaw of a dog, and they walked on the other three digits.

The head, however, is obviously where the business end of a carnivorous animal typically lies. Most theropods had a series of curved and serrated teeth in the jaws that would provide a cutting action on whatever was in the mouth when it closed. The teeth did vary quite substantially in size, shape and number in various lineages in accordance with different prey and feeding styles. Most notably, a number of theropods followed the sauropodomorphs and ornithischians, and switched to a vegetarian (or at least omnivorous) diet, eschewing killer teeth for those that could shear plants, or losing the teeth entirely and adding beaks to the snout.

Most non-avian theropods had feathers of various kinds, but actually most, if not all, saurischians had another more interesting avian feature, namely air sacs. Extensions of the lungs fed out and into both the body cavity and the bones themselves, making them pneumatic. Pterosaurs actually have the same system, and it's not impossible that this trait appeared before the origin of the dinosaurs and was then lost by the ornithischians, rather than being a special feature of the saurischians. This feature, however, is a major link between the birds and the dinosaurs, and indeed it unites the theropods and sauropodomorphs into the saurischians. It also had big implications for things like how we estimate the mass of these animals: if a lot of them were air by volume, they were probably rather less heavy than we previously imagined.

In many cases, therefore, identifying the bones or teeth of a theropod when out in the field or rummaging through museum collections is relatively simple. A curved, serrated tooth is probably that of a theropod, a lizard-type pelvis or hollow, pneumatic bone is clearly that of a saurischian, and if it's small it's more likely to be a theropod than a sauropodomorph. Any kind of clawed hand will belong to a theropod, and while not discussed here, many theropods have distinctive vertebrae, too. Identifying a theropod, then, should be easy, but how can you tell if what you have is a tyrannosaur as opposed to any of the other numerous lineages of theropod?

## The tyrants

The tyrannosaurs are, of course, distinctive in their own right. Although the general anatomy and form of the later tyrannosaurs is easily recognisable with a little knowledge, the earlier tyrannosauroids are rather less different from the other theropod lineages that they branched from. Give it a few million years and their respective evolutionary trajectories will have taken them away from each other, and their descendants will look quite distinct.

Thus the earlier tyrannosaurs cannot be distinguished from the other theropods to which they are most closely related (the compsognathids, and other assorted basal coelurosaurs) simply by major features such as the huge heads and small hands of the tyrannosaurines; instead, you need to delve a little deeper into the anatomical details of the tyrannosaurs. Our modern understanding of tyrannosaur taxonomy is massively influenced by ever-newer methods and discoveries, and this is the reason why some species were originally overlooked as tyrannosaurs, and why there was confusion over where tyrannosaurs fitted in the theropod tree. Simple definitions of suggested clades resulted in the equivalents of the pachyderms, perhaps best exemplified by the putative group 'Carnosauria'.

In the absence of many smaller dinosaurs and transitional forms (and a lack of detailed examination of the anatomy), all the big carnivorous theropods were originally lumped together in one group called the carnosaurs.[3] This naturally included *Tyrannosaurus*, but also other beasts such as *Allosaurus*, which were large-bodied carnivores, too. Certainly, various members of this group did share features in common, such as their large size, but we now know that most of them should be separated off. Similarly, the definition of the Carnosauria would exclude a number of taxa that we now know to be tyrannosaurs. For example, the small *Proceratosaurus* was originally (as its name suggests) thought to be a forerunner of *Ceratosaurus*, with both bearing large crests on their snouts; it was later considered to fall somewhere within an earlier branch of theropod evolution, but is now widely regarded as an early and rather small tyrannosaur.

*Proceratosaurus* has been reassigned because it bears the key characters that we now associate with tyrant dinosaurs as a whole. The tyrannosauroids are diagnosed in part by having a rather short premaxilla and also heterodonty – that is, the teeth in the jaws are not uniform in shape (or close to it), as they are in many reptiles (including theropods), but instead show different forms. More specifically, the teeth in the premaxillae are rather sub-circular or even D-shaped in cross-section (with the flat of the D facing back into the mouth and the arc facing forwards). This shape is actually similar to the incisors of many carnivorous mammals, and is quite different from the more normal flattened shape of the rest of the teeth in the tyrannosaur jaw (even these teeth are typically more robust in tyrannosaurs than in other theropods).

One other key early feature of tyrannosauroids is that the nasal bones are fused together in adult animals. This pair of long, thin bones runs along the midline of the skull behind the premaxillae, and between the left and right maxillae, and helps to make up part of the top of the snout and to outline the back of the nares. In tyrannosaurs of all kinds the nasals are fused together into a single unit, and as this is a feature that strengthens the skull as a whole during biting, it suggests that it was a major part of the early evolution of the group, as well as providing a good way of identifying tyrannosaurs. Similarly, all later tyrannosaurids have proportionally long metatarsals (a feature that reappears in birds as well as troodontids and ornithomimosaurs, but is still useful to help define the clade here). This feature links to a different locomotory pattern from other animals, and again suggests an early adoption of this feature by the group.[4]

These characters therefore help us to work out what is and what is not a tyrannosaur. Note that while originally such features would have been ascribed under a basic taxonomic system (basically people noted a number of things that had several features in common, and named a group based on that), now these characters are established primarily through phylogenetics. That is, we run phylogenetic analyses using *all* the data, which enables us to see which features unite certain groups. These are then useful indicators for establishing the relationships of new finds without having to go through a full analysis. While it's always possible that you've found some quirky new animal, if you find a partial skull with D-shaped premaxillary teeth and fused nasals, you can be pretty certain it is a tyrannosaur without having to run it through a cladistic analysis. Similarly, bear in mind that these characters are used to help define the group but are not absolutes: birds are saurischians despite the shapes of their pelvises. The key to defining which group an organism belongs to lies in its evolutionary ancestry, not its appearance. Therefore, although obviously it's the latter that helps determine the former, evolution can change features quite considerably given enough time.

Moving on, the proceratosaurids represent the earliest branch from the tyrannosauroid lineage. They can be identified by the greatly enlarged nares and the nasals being expanded into a large crest,[5] which gives the skull a very distinctive profile. It was once suggested that the procertosaurid *Guanlong* was in fact synonymous with the odd Chinese allosauroid *Monolophosaurus*, and the two do share these features in common (as indeed does *Proceratosaurus*), but *Monolophosaurus* is quite devoid of tyrannosaur characters and *Guanlong* has lots of them.

As we get to the more familiar and larger-bodied animals, the tyrannosaurids, more familiar features become apparent and the overall shape that most people might identify as

*Fig. 5a Representative skeletons of tyrannosaurs from various major clades. Top to bottom: The proceratosaurid* Guanlong, *the tyrannosauroid* Stokesosaurus, *the albertosaurine* Gorgosaurus, *and the tyrannosaurine* Teratophoneus.

tyrannosaurian starts to become apparent. The D-shaped front teeth and fused nasals are still there, but we also now see key features such as the maxillary and dentary teeth becoming more robust, the nasals narrowing towards the back of the skull, and the claws on the hand being the same size as each other (unlike in early tyrannosaurs and other theropods, where the claw on the first finger is rather larger than the others).[6]

The pelvis is modified, too; looking down from above, the two blades of the ilium lie almost together alongside the neural spines of the sacrum, and the whole thing looks as though someone grabbed it and pinched everything together. While taken to an extreme in the tyrannosaurids, *Guanlong* also has something similar, so this feature may have appeared, or at least started to evolve, earlier in the line. The ischium had changed in the tyrannosaurids, too; although the pubic boot remains large, most dinosaurs have an expansion to the end of the ischium, or at least it has a rounded end, but the tyrannosaurids have lost this feature at this juncture, and the ischium tapers to a point.

Finally, the tibiae have elongated and the metatarsals are even longer than in the early tyrannosauroids. The specialised pinched bone in the foot – the arctometatarsal – is also diagnostic of tyrannosaurids, and while this appears independently in other lineages (such as the oviraptorosaurs), within the tyrannosaurs it is unique to tyrannosaurids, and helps to distinguish them from earlier forms. Most notably, the hand now has only two fingers: although a third metacarpal survives as a splint of bone in many specimens (including a couple of *Tyrannosaurus*), there are only two functional digits in any tyrannosaurid.[7]

The tyrannosaurids contain two major groups, the alberto-saurines and the tyrannosaurines, though recently a third has been proposed to lie inside the latter: the alioramins. These tyrannosaurs have rather elongate and low skulls, lack the 'giant' heads of their relatives and also have rather thin teeth compared with those of other tyrannosaurines.[8] The albertosaurines have a proportionally lower skull than the more typical tyrannosaurines, but it is taller and larger than that of the alioramins, and they have longer lower legs, metatarsals and toes than the other tyrannosaurines.[9] In albertosaurines, a number of details in the skull, such as the nasals no longer contribute to any part of the edge on the antorbital fenestra (that is, their shape and position have changed relative to the maxillae). At a basic level you could informally separate the albertosaurines as being those tyrannosaurids that lie intermediate in form between the other two clades: they are not as slender as the alioramins, but not as bulky as the tyrannosaurines.

In addition to having generally larger heads and overall more robust proportions in the adults when compared to albertosaurines, let alone other theropods, the tyrannosaurines have a notably curved lower edge to the maxilla, giving the jawline a rather distinctive profile when the mouth is open. The skull continues to change over time, with the antorbital fenestra becoming taller than long, and the shape of the orbit changes: in the later forms at least becoming constricted in the middle, giving it something of a figure-of-eight shape (with the actual eyeball sitting in the top part of the '8'). Once again, the pelvis is rather key, with the ilium now being longer than the femur, and the pubic boot really increasing to mark it out as an especially modified part of the hips.

Being armed with such knowledge makes life relatively simple when sorting through tyrannosaur fossils. It also immediately reveals some transitions and changes that mark out how these animals were evolving: with the ever-increasing pubic boot and lengthening metatarsals, the changing head shape, hand reduction, nasal modifications and so on. Some of these changes are obviously linked to major functions like locomotion, but others are perhaps less obvious and may simply be a case of random changes becoming evolutionarily fixed in the history of their respective lineages. In any case, they provide a beginning of an understanding of how the group changed over time, and the biological implications associated with it.

However, such a discussion is based on the fact that we can (relatively) confidently place individual species of tyrannosaur into clades and determine the relationships between them. Knowing who is related to whom, we can then see what they share in common and how they differ, and thus determine the changes that have happened. For this, we need to delve into tyrannosaur relationships go beyond the characters described above that allow us to identify these groups, and see why we are able to identify these traits as diagnostic. We will look at this in detail in Chapter 4, but first we need to use these characters to sort out our tyrannosaurs into different species.

# CHAPTER THREE

# Tyrannosaur Species

Just how many species of tyrannosaur are there? It's far from a simple question, but a critical one to answer if we are to study tyrannosaurs' evolutionary history. Dozens of tyrannosaur species have been described at various times over the last century or so, but not all of them are considered valid species by most researchers; as we've seen, names like *Dynamosaurus* and *Deinodon* are now no longer in common use, and others like *Shanshanosaurus* are controversial. Numerous fragments of bones, odd teeth and the like have been named as belonging to new genera and species at various times, and while in most cases these do belong to tyrannosaurs, they have no defining characteristics that mark them out as genuinely distinct lineages. For example, an incomplete scapula (shoulder-blade) was named in 1958 as belonging to a new genus of tyrannosaurine called *Chingkankousaurus*, but detailed analysis of this bone by Steve Brusatte and colleagues (2013) shows that it has nothing that can allow it to be distinguished from the bones of any other large tyrannosaur.[10] The name is therefore a '*nomen dubium*' (literally, a dubious name), and is not then considered valid. Other animals are more controversial, and there can be fierce disagreements between researchers as to whether or not they have valid names. Principal among these is *Nanotyrannus*, considered by some to be a very interesting dwarf tyrannosaur that lived alongside *Tyrannosaurus*; most researchers think that the specimens assigned to *Nanotyrannus* merely belong to juvenile specimens of *Tyrannosaurus* (a subject returned to later).

Leaving aside the questionable forms and invalid names, there are currently 29 known species of tyrant dinosaur that are near-universally agreed upon. The number is growing rapidly (at least one new species has been named a year for the last decade or so), and with a few other specimens knocking around that probably also represent new species, but have yet to be formally named by researchers, this total is likely to go up before this book is even published*. The number has been boosted in recent years in part through new discoveries like those of *Yutyrannus* and a second species of *Alioramus* (*A. altai*), but also through the recognition that some animals previously thought to belong to other groups are in fact tyrannosaurs, the most obvious being *Proceratosaurus*. Until relatively recently, this was regarded as belonging to a group called the ceratosaurs, as might be guessed from the name, but new studies, in particular those led by Oliver Rauhut in 2010, suggested that it was in fact an early tyrannosaur. Such changes are not uncommon in palaeontology, based on data from new skeletons or more detailed analyses of existing material, and because of this additional research and a flurry of increased discoveries the number of tyrannosaurs is increasing.

The total of a couple of dozen species may not seem like that many, given the diversity of, say, modern dog and cat species alive right now, let alone groups that have existed for more than 100 million years, but we are still getting to grips with dinosaur diversity, and there are undoubtedly many more species to discover. In addition, the tyrannosaurs suffer from being relatively large carnivores, which are fundamentally rare components of ecosystems. Go to Africa and you may see only one or two lions for every hundred zebra or antelope, and since there are also hyenas, leopards and other competing predators out there, lions make up only a small number of the total individual mammals in a given region. The situation is inevitably more complex than this in

*This sentence turned out to be rather more prophetic than I'd anticipated – I've changed it three times since I first wrote it.

terms of fossil finds, but carnivores, and especially big carnivores, are simply rare, so are harder for palaeontologists to find than more common animals.

Compared with other groups of large carnivorous dinosaur, the tyrannosaurs are quite diverse (Fig. 5a), and the numbers we have reflect that diversity. Given the extreme interest in them, and the fact that many are from North America, they have in fact been the subject of rather more research than groups that are predominantly African or South American. For example, there are currently around 20 species known from the allosauroids and 12 from the ceratosauroids, each of which is dominated by large-bodied carnivores, while the famous spinosaurs are known from less than 10 species, despite a distribution that covers Africa, Asia, South America and Europe over at least 30 million years (and most of them are known from only a few fragments). This level of research doesn't inflate the number of tyrannosaurs, but it does mean that the other groups might be a bit behind when it comes to the numbers of species found and named, and overall the various clades might be more comparable in terms of numbers than they currently appear.

Below is a list of all currently known and valid tyrannosaur genera and species. Doubtless some researchers will disagree with the list (a few, for example, prefer *Tarbosaurus bataar* to be considered *Tyrannosaurus bataar*), and although subsequent studies will certainly add to it, a few other species may yet slip off it if they are identified as belonging to an existing species, as our understanding of their anatomy improves and new finds are made. In fact, around 30 per cent of all newly named dinosaur species are eventually struck off the register, as were *Chingkankousaurus* and *Deinodon*. In part this can be attributed to the overly liberal naming practices of the past, but some new species continue to be named based on poorly constrained characters and data that is too limited.

**Table 1** All known and currently valid tyrannosauroids in their appropriate clades where known (some are uncertain, so are placed in the broadest clade to which they definitely belong), then listed alphabetically within each clade.

| Clade | Genus | Species | Location |
|---|---|---|---|
| Tyrannosauroidea | *Alectrosaurus* | *olseni* | Mongolia |
| | *Appalachiosaurus* | *montgomeriensis* | Alabama, USA |
| | *Aviatyrannis* | *jurassica* | Portugal |
| | *Dilong* | *paradoxus* | Eastern China |
| | *Dryptosaurus* | *aquilunguis* | US |
| | *Eotyrannus* | *lengi* | UK |
| | *Juratyrant* | *langhami* | UK |
| | *Santanaraptor* | *placidus* | Brazil |
| | *Stokesosaurus* | *clevelandi* | UK & US |
| | *Xiongguanlong* | *baimoensis* | Northeastern China |
| | *Raptorex* | *kriegsteini* | Mongolia |
| | *Yutyrannus* | *huali* | Eastern China |
| Proceratosauridae | *Guanlong* | *wucaii* | Western China |
| | *Kileskus* | *aristotocus* | Russia |
| | *Proceratosaurus* | *bradleyi* | UK |
| | *Sinotyrannus* | *kazouensis* | Eastern China |
| Albertosaurinae | *Albertosaurus* | *sarcophagus* | Alberta, Canada |
| | *Gorgosaurus* | *libratus* | Alberta, Canada |
| Tyrannosaurinae | *Alioramus* | *altai* | Mongolia |
| | *Alioramus* | *remotus* | Mongolia |
| | *Daspletosaurus* | *torosus* | Alberta, Canada |
| | *Lythronax* | *argestes* | Utah, US |
| | *Nanuqsaurus* | *hoglundi* | Alaska, US |
| | *Qianzhousaurus* | *sinensis* | Southern China |
| | *Tarbosaurus* | *bataar* | Mongolia |
| | *Teratophoneus* | *curriei* | Utah, US |
| | *Tyrannosaurus* | *rex* | Western North America |
| | *Zhuchengtyrannus* | *magnus* | Eastern China |

## What makes a species a species?

Identifying and naming a new species is naturally a big deal, and knowing how many species there are and which is which is very important, hence the focus (and more than occasional heated arguments) about the identities of species. Correctly identifying species, even in living organisms, is rather difficult, and there are plenty of disagreements among taxonomists working on extant animals, as well as palaeontologists looking at fossils. The often-used definition of a species that it is 'a group of organisms capable of interbreeding and producing fertile offspring' is just one of many definitions in use. This has its limitations: it's not a lot of use on fossils for a start, and it is hardly practical to keep trying to see if any given pair of animals can interbreed. As a result, biologists use a variety of 'species concepts' to identify species, be it from their genes, behaviour, interbreeding or other factors. In the case of palaeontology we are reliant on what is called the 'morphological species concept': we largely have only morphology (anatomy, and more specifically the bones and teeth) to go on, so base our definitions on this. Actually it's no better or worse than many other methods, and of course many living species can be told apart by their anatomy just as well as by any other features. Looking out of the window, you probably identify most birds in your garden by their differences in plumage, but their skeletons also look different, and indeed they behave differently, have different genes and only tend to breed with their nearest relatives. It is not ideal to be so reliant on anatomy, but it is no better or worse than many other available options, and mostly these different concepts do line up pretty well.

Still, it can be difficult to identify a species correctly from the bones alone, and when faced with only half a skeleton or less it can be hard to say whether or not something really is new. *Chingkankousaurus* could be a new tyrannosaur, but as it cannot be told apart from animals we have already named, for now it must remain of uncertain identity, just as a single brown feather in the garden could have come from a new species, but could easily be from a sparrow or a wren. This approach may seem overly conservative, but things like the teeth of tyrannosaurs are extremely common (while visiting Alberta in Canada with two colleagues we found several in one morning), and while easy to identify as belonging to the *group*, beyond that there's very little to say, and it is surely best not to name a new species every time a new tooth turns up that is a little bigger or smaller, or has a few more serrations, than the other teeth.

Additional complications come from the fact that, like most vertebrates, dinosaurs change in shape as well as size as they grow. A hatchling *Tyrannosaurus* looked rather different in many details from a half-grown animal, and both were different again from a full adult. It is important, then, to make sure that names are pegged onto adult animals or, if only juveniles are known, to be confident that their characteristics are still diagnostic. For example, many mammals change colour as they get older, and the milk teeth of babies might be different from those of adults, so using colour or tooth pattern to try and separate an adult of one species from a juvenile of another is not a good plan. In particular, babies of all species tend to have very big heads and huge eyes, so these are real areas to avoid – though any fossil find with disproportionally large eyes is immediately a good candidate to be a very young animal, so is of great interest. On the other hand, things like the number of digits in the hand or foot, or the number of bones in the neck, probably don't change, so differences here tend to represent real distinctions between species. Things become clearer still when you have both adults and juveniles of a single species, as is the case, for example, for *Tarbosaurus*.

Even at the same age, not all members of a species look exactly alike (Fig. 5b). Some people are taller than others and may have brown or black hair, and grey or green eyes, or may have large ears or a Roman nose, but there are deeper differences in the skeleton and teeth, too. It's not uncommon to find people with one more or one fewer bone in the coccyx, with or without wisdom teeth, with depressed sterna, or with no clavicles, and other variations. Some of these

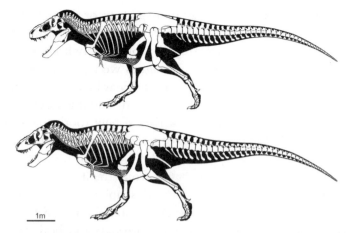

*Fig. 5b Not every individual of a species is exactly alike. Here are two different specimens of adult* Tyrannosaurus *and they are not quite the same in terms of size or details of the skeleton like the proportions of the skull and size of the pubis.*

conditions are much rarer than others (though they may be common in some populations), but they are all part of the pattern of human variation. Speaking personally, I have a rather odd-shaped skull that is quite apparent when I cut my hair short, but I sincerely hope that I'm not considered separate from the human species as a result. We do, however, need to take account of these differences within species (known as intraspecific variation) and not split hairs: just because your new skull has one more tooth than all other specimens of *Tyrannosaurus rex*, it doesn't mean you have a new species.

As it happens, some bits of animals are incredibly variable and thus need to be treated more cautiously. This applies especially to tails, where the number of bones varies considerably across a great many vertebrates, and this is also true of dinosaurs.[11] Tail length can vary a great deal between even very closely related dinosaur species, and within species there can be some big differences too. This means that for animals with no tail known it can be hard to predict how long it might have been, and even those animals with a lot of tail preserved might have had quite a bit more than we would normally expect. This is one reason why I generally shy away from the traditional 'total' lengths of dinosaurs as an estimate of their size: based on one single measurement, an animal with a long tail looks as though it's a lot 'bigger' than one with a short tail. A bobcat is not much longer than a domestic cat, but one is all cat and the other is all tail. Coupled with how few dinosaurs we have decent tails for, total length becomes pure guesswork. However, for the tyrannosaurs this probably isn't too much of a problem, as they were all built very similarly, but it's impossible to rule out a find of a very long- or short-tailed animal that would rather skew things, and so while I do give estimates of the total length in this book, beware that these may be under or overestimates for some species.

Finally, there can also be differences between males and females known as sexual dimorphism (literally 'two different shapes'). Again, while some of these can be quite dramatic in the animal kingdom (few should mistake a peacock for a peahen, or a red deer stag for a doe), even in animals with limited differences between the sexes these can produce clear patterns in the bones, both obvious and slight. However, sorting out such differences in dinosaurs is rather harder, especially when often we have only one or two specimens of a given species that might have lived hundreds of thousands of years apart.

Despite extensive work, there are few dinosaur species for which we have even a little confidence of being able to tell males from females. In part this is down to a narrow set of data: many species are known from only a single specimen or just a few, so there's not a lot to work from. Even in those for which we have dozens of complete skeletons the differences seem to be pretty minor, so it is not evident if there are enough clear and consistent differences to separate out the specimens into two groups, or to group them in more of a continuum. Even if there are two groups that can be identified, which would be male and which female isn't known either. A few differences have been suggested in the past that might give clues to the identity of a skeleton's sex, but these have proved to be erroneous. For example, it was suggested that females would have one less chevron in the base of the tail than males, to better facilitate egg laying, and while some skeletons of dinosaurs do seem to have fewer chevrons than others of the same species, there's no indication that this links to any other differences, and indeed this pattern is not linked to sex within living reptiles, so it seems to offer little help. In short, if someone claimed that a specimen of a given dinosaur, or indeed tyrannosaur, was a male or female based on its overall appearance, I would be reaching for the salt cellar to take a large pinch the vast majority of the time. That said, this doesn't mean that females cannot be identified in some cases, due to a wonderful quirk of reptile and bird reproductive biology that leaves traces in the bones.

Animals that lay hard-shelled eggs need a supply of calcium, and quite a lot of it at short notice, when they are laying eggs, to help make up the shell. Therefore in the run-up to and during the breeding season, female birds and reptiles lay down a special kind of bone on some parts of their skeletons. This material is called medullary bone, and it is set so that it can be broken down easily to dump its calcium into the bloodstream and help make eggshells. Take a chunk from the femur, grind it down until it is wafer-thin and shove it under a microscope and, if present, the characteristic form of medullary bone is easy to see, and hard to mistake for anything else. If you have this, the animal must have been a female, though of course its absence could simply point to a female out of the breeding season (or even an immature animal) rather than to it being a male. In theory you might be able to identify a male if several animals were found that were preserved together at the same time: if some were carrying medullary bone and others not, then the breeding season was probably in full swing and the animals lacking it were probably male. If these could be shown to have some consistent difference in skeletal morphology, too, the case would be quite solid. Medullary bone has allowed palaeontologists to identify at least a few female dinosaurs with confidence, but unfortunately there's little on tyrannosaurs so far.

All of this adds up to a tricky question of accurately identifying whether or not any given specimen really represents a new genus or species, or simply a slightly odd version of an existing named taxon. Not all researchers will agree as to how different an animal should be in order to require a new name, and there can be (and are) some embittered battles about identity. What is rarely in doubt is the actual differences: no one would dispute that *Raptorex* and *Nanotyrannus* look rather different from *Tarbosaurus* and *Tyrannosaurus* respectively, but whether or not these are genuinely different species, or merely juveniles or rather odd-looking individuals, is the problem. When arguments settle on small subtleties and specimens are incomplete, things can become especially difficult.

## Naming a new species

For all that, however, there is something of a consensus in the tyrannosaur world. The odd contrarian aside, most researchers agree on a pretty similar set of species, with the same specimens being assigned to them. At the same time, people are quite willing to look at new data or new analyses, and some arguments can be brought forwards to overturn old ideas. A lot of the disagreements centre around the same set of people arguing over the same set of specimens, so actually this doesn't have too much bearing on wider studies, provided you are aware that the work of tyrannosaurologist X will always count *Nanotyrannus* as an additional genus, and tyrannosaurologist Y will not.

Although disagreements can occur even when there are whole skeletons to play with, some surprisingly small bits can be enough to be identifued confidently as belonging to a tyrannosaur, or even as a new species. In 2010, I was the lead author on a paper that named a new, large tyrannosaurine – *Zhuchengtyrannus magnus* – from eastern China.[12] Although quite a few tyrannosaur bones were recovered from the quarry, these included parts of skulls of two rather different tyrannosaurids, making it impossible to know which parts might belong to which head without a whole animal as a guide. As a result, even though there were femora, metatarsals, teeth, dorsal vertebrae and even a couple of ribs known, *Zhuchengtyrannus* was ultimately named on the basis of only an incomplete maxilla and a dentary. Not only that, but nothing in the dentary was unique, so in fact the whole identification of this as a new genus was based on the maxilla. However, we were able to identify a number of unique features of this not seen in any other tyrannosaur, and especially not in *Tarbosaurus* (the nearest tyrannosaur in time, space and size, making it the most likely candidate if this was not new). The maxilla is an especially important diagnostic element, in that it contains a large number of detailed anatomical characters that are used to identify species and relationships between them. This probably would not have been possible with a caudal vertebra or a humerus, but it does show how little material can be needed to correctly identify a new animal.

There are formal hoops to go through in order to name a new species, and since this can take time, it's not that uncommon for people to identify specimens that probably represent new species, but that have yet to be named. Look through the scientific literature and as far back as 1992 there are mentions of the 'Two Medicine Formation' tyrannosaur. Yes, it does come from the Canadian 'Two Medicine Formation' (a geological series in Alberta that extends into Montana in the US), and no, it's not clear why it has been sitting around for 20 years without having a name stuck on it. Some other specimens fall into a bit of a tricky middle ground: not enough is preserved to be able to tell them apart from other species, but they are almost certainly new. For example, in 2010 Roger Benson and colleagues described a partial pubis from south-east Australia[16]. There is not much of it, but enough to say that it belonged to a basal tyrannosaur with some confidence, and given the distance between Australia and the next-nearest tyrannosaur fossils it is likely that this represents an as-yet unidentified species.

Fortunately, it's not all bone fragments and teeth, and a number of tyrannosaurs are known from huge amounts of superb material. *Dilong* is known from a near-complete and very well-preserved specimen, for example, and *Guanlong* is known from at least three specimens, two of which are near complete and one of which is a juvenile. Each of the Canadian trio of *Albertosaurus*, *Daspletosaurus* and *Gorgosaurus* is known from multiple, beautifully preserved specimens, some of which are complete almost down to the last tooth, rib and chevron, and *Tyrannosaurus* itself is not to be left out, with a good number of largely complete fossils.

# CHAPTER FOUR

# Tyrannosaur Relationships

Despite all the arguments and issues surrounding the identification and naming of species, we do have a good idea of the number of tyrannosaur species (and genera) – at least at the moment – and we have discussed the characters that unite various groups. Knowing which species is which, and which specimen belongs to which species, is a very important jumping-off point, but in order to really understand the evolution of a group of animals, we need to know how the tyrannosaur family tree branched to produce the various lineages and species.

A quick look at some of the features various genera have in common gives you an idea of which might be closer to each other, and which might be further apart. This is largely the approach taken by traditional taxonomists: grouping things together based on a few shared characteristics, but such a method provides limited resolution and risks the already mentioned 'pachyderm' problem of picking the wrong traits. Certainly, *Tyrannosaurus*, *Tarbosaurus* and *Zhuchengtyrannus* have enough in common for us to consider them to be close relatives (they are bigger, and have fewer maxillary teeth in a taller skull, than any other tyrannosaurs), but it is both useful and interesting to know exactly who might be closer to whom, and that requires a lot more detail.

The techniques used to produce family trees of organisms and suggest how they might be related is called phylogenetics, and the resultant 'trees' are termed phylogenies. Their creation essentially boils down to an assessment of all the available data for a species and comparison with the data for all the other species, to determine which species have the most features in common. Those that are closest relatives will have the most details in common, since they have the most amount of shared evolutionary history, and have changed the least amount after they diverged from their ancestors, while those that are less close relatives will have less in common.

Early phylogenies were completed by hand: a few dozen characteristics would be compared between a handful of species or lineages, and a tree would be assembled from this data. This method had two obvious problems: it was horribly time-consuming, and the amount of information that could be considered was moderate at best. Manually comparing even 20 or 30 characters between a dozen or so species could take all day, and of course there was always the risk of human error. This was a clear improvement on the old system of simply choosing a few almost arbitrary characters and considering them the most important, but going from, say, half a dozen characters or so to a couple of dozen really wasn't a colossal leap forwards. Although you could now provide greater resolution and potentially produce an exact relationship between any three or four species (unlike before), this was inevitably a very slow process without some form of automation.

Doing manual assemblies of trees was how I originally learnt to produce phylogenies as an undergraduate, way back in the not-actually-that-long-ago year of 1997. It was rather fun and it really taught me to get an understanding of the underlying data, and how changes or additions to the data could affect the tree at the end. But as a process it was inevitably going to be superseded by computing.

This took longer than you might expect. Although computers were pretty ubiquitous in science by this time, and phylogenetics software was available, the numbers being handled for even small trees were absolutely astonishing. As in the problem of the Maharaja's rice (a supposed gift of one grain of rice for the first square on a chess board, two for the second and so on), the numbers get very large, very quickly. For three species there are only three possible outcomes: A and B closest, then C, B and C closest, then A, and finally A and C together, then B. However, for four species there are 15 combinations (A and B, then C, then D, A and B, then D, then C, A and B, then C and D as a second pair, and so on), but five species give 105 possible trees, and when you have only 10 species there are already 34.5 million possibilities. When you get into the dozens or hundreds of species, even working out the number of possibilities becomes tricky. As a result, it took a while before the processing power of the average desktop PC could handle such things, and before the software was refined enough to be able to exclude large numbers of definitely incorrect trees quickly, before exhaustively searching through a relatively small number to try and find the best solution.

Nowadays it is relatively common to find analyses of fossil vertebrate groups featuring dozens of species and assessing hundreds of characteristics. Recent analyses of tyrannosaurs have included pretty much every known species, and more than 300 characters have been included, covering everything from the shape of the maxilla and the denticles (small serrations) on the teeth, right through to the attachment sites for muscles on the pelvis, the shape of the ankle and the structure of the braincase. In short, a colossal amount of data has been collected and processed, and almost every bit of detail that can reasonably be extracted has been assessed and collated. Putting all of that together means that the results of the analysis are the best that we can reasonably expect to produce under the circumstances (missing data, for example from incomplete skeletons, can cause problems, or mean that we can't always include everything, and of course since juveniles may not look like adults, these must be treated with caution).

## Putting the tyrannosaurs in their (evolutionary) place

Let's start with the dinosaurs as a whole, and where various major players fit in the tree of life, or at least within the more recent radiations of the reptiles. Tyrannosaurs are just one group (or 'clade': a better evolutionary term) of theropod, and it is important to understand how they relate to other forms. This is important since it is key to finding out which traits are uniquely tyrannosaurian and which they have inherited from their ancestors. As can be seen from the branching pattern of the tree, the theropods are actually closest to the sauropodomorphs: the group that includes all of the huge and famous long-necked giants such as *Diplodocus* and *Apatosaurus*. All other dinosaurs, including the horned dinosaurs, armoured dinosaurs and hadrosaurs are in the ornithischian clade (Fig 6). Dinosaurs are actually rather close relatives of the non-dinosaurian pterosaurs (often mistakenly called pterodactyls, and also occasionally

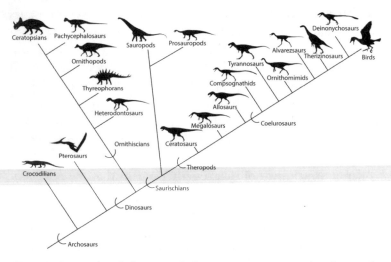

*Fig. 6 A simple phylogeny of the major groups in the theropods, showing some of the other dinosaur groups and their nearest relatives the crocodiles and pterosaurs. Huge amounts of diversity are masked here (there are around 250 species in the sauropods for example) but it gives an idea of their relationships and shows the position of the tyrannosaurs. Silhouettes are representative and not to scale and the lengths of the lines are not representative of the lengths of time these lineages were alive.*

mistakenly considered bird ancestors), and also of croco-dilians: a group now much diminished, but formerly highly diverse. Birds are very much dinosaurs, and indeed derive from a group of bird-like dinosaurs called the deinonychosaurs.

In all cases these trees are a little simplified: not everything sits in neat groups, and there are some single species that sit between those shown here, and others whose exact position is uncertain. This should, however, provide a good approximation of the major consensus of palaeontologists on the positions of these groups. The silhouettes here and used below give a general idea of the overall shapes of the animals within their respective clades, though in many cases this massively understates the diversity in form and size (after all, birds include peacocks, parrots, kiwis, penguins, ostriches, hummingbirds and albatrosses).

Within the theropods, the tyrannosaurs take a rather midway position. They were not the first group to evolve, but nor were they the last. The names of clades are often taken from the first or best-known member of a group (tyrannosaurs obviously come from *Tyrannosaurus*, allosaurs from *Allosaurus* and so on), though others are more cryptic to the uninitiated: the legendary *Velociraptor* is a dromaeosaurid, for example, and few are familiar with the therizinosauroids. As can be seen, the nearest relatives of the tyrannosaurs are the compsognathids and ornithomimosaurs, and tyrannosaurs are close to the base of the group called coelurosaurs, meaning 'hollow reptiles'. Although hollow bones are now known to have existed well before this point, the name comes from the plethora of pneumatic, bird-like bones seen throughout the members of this great clade.

Zooming in, as it were, here is the 'industry standard' phylogeny of the tyrannosaurs (Fig. 7a). There are three major divisions to denote various clades that therefore represent the evolutionary divergences of the respective clades. Early on the proceratosaurids branch off, and later within the tyrannosaurids there are the alioramins, alberto-saurines and tyrannosaurines. Despite some uncertainty in the ages of the rocks that have yielded various tyrannosaurs,

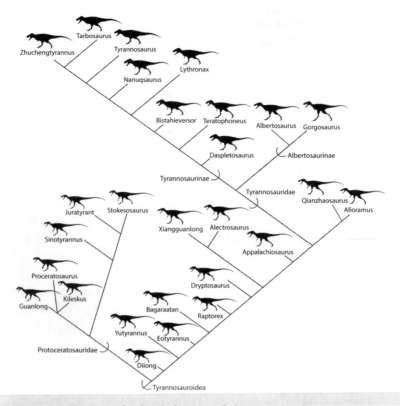

*Fig. 7a The 'standard' phylogeny of the tyrannosaurs. Not every known genus can be placed and some of the relationships shown here are uncertain. Silhouettes are not to scale and the lengths of the lines are not representative of the lengths of time these lineages were alive.*

overall the evolutionary pattern of this phylogeny matches the dates that these lineages appear in the fossil record. That is, the lineages that we hypothesise arose later do appear in more recent (closer to the present) rocks. Incidentally, this is one of the reasons to think that cladistics produces broadly accurate results. Studies have shown that the branching patterns from our trees of all kinds of fossil organism are, on average, a good match for the ages of the first appearances of various fossils, giving us confidence in the analyses.

Inevitably, however, there are disagreements over the exact arrangement of the various lineages in various analyses. Different researchers use different combinations of characters and even species for their studies, to produce subtly different trees. For example, some trees have suggested that *Bistahieversor* might be placed within the tyrannosaurines, a rather different location from the one shown here. Overall, though, things have generally been pretty stable – until very recently, anyway.

First of all, Argentinian palaeontologist Fernando and colleagues suggested in 2012 that the tyrannosaurs might be rather larger as a group than had been previously realised.[13] They suggested that a number of additional species were either very closely related to, or even in, the tyrannosauroids. This set of species includes the enigmatic *Megaraptor* from the Cretaceous of Argentina, and *Fukuiraptor* from Japan: animals that have typically been considered to be part of the allosauroids, and indeed, relatively derived members of this group. There are some anatomical details that support this hypothesis, but it does lie in rather stark contrast to previous analyses of both the allosauroids and tyrants. It's certainly not impossible, but for now we'll set this aside.

The second big shift hypothesised came with the arrival of *Lythronax* in 2013, and with this new genus came a new analysis of the evolution of the tyrannosaurs. In the paper describing and naming this animal, author Mark Loewen and colleagues presented a tree that was bordering on radical in how different it was from the previous interpretations.[14] In their work, numerous taxa moved from their 'normal' positions, and the organisation of the tyrannosaurs was rather different. This has some rather important implications, not least for how the various tyrannosaurs, and especially the tyrannosaurines, may have dispersed and moved between the continents. This subject is explored further in the section on tyrannosaurs in time and space, but the evolutionary implications are dealt with here.

Looking at the Loewen phylogeny here, we can see that there has been quite a reshuffling between groups (Fig. 7b).

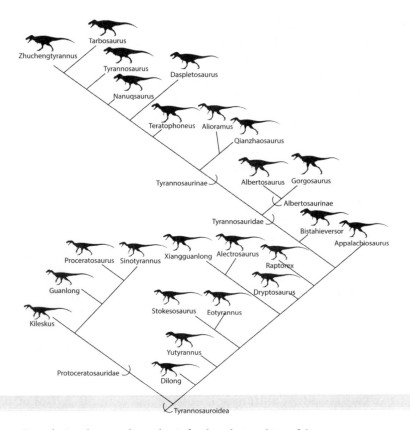

*Fig. 7b An alternate hypothesis for the relationships of the tyrannosaurs. Note especially the change in the position of the alioramins and* Bistahieversor. *Silhouettes are not to scale and the lengths of the lines are not representative of the lengths of time these lineages were alive.*

The major lineages are all still there, and most of their representatives are still in place (for example, *Tyrannosaurus* and *Daspletosaurus* are still tyrannosaurines, *Albertosaurus* and *Gorgosaurus* are still albertosaurines, *Guanlong* is still a proceratosaurid), but other lineages have changed. For example, *Bistahieversor* is now a tyrannosaurine and *Alioramus* is no longer a member of the tyrannosaurines, while *Teratophoneus* is rather closer to *Tyrannosaurus* than it used to be.

Naturally, it is interesting to know how we think the various genera and species are related to one another, but phylogenies also give us a way of examining how certain traits have evolved, when they occurred, the rate at which they have changed, and issues such as which features the groups most have in common. This allows us to evaluate the evolution of lineages, but with a different arrangement of genera our interpretations of these patterns will also alter. One major feature we come back to later is the length of the arm of various tyrannosaurs, but it's instructive to take this as a brief example.

While *Tyrannosaurus* and other tyrannosaurines famously have very small arms and hands, in early forms of these animals they were of comparable size to those of many other theropods. There is actually a fairly consistent pattern of reduction from the first tyrants towards the last tyrannosaurines, but if, for example, a new analysis suggested that *Tarbosaurus* was in fact a proceratosaurid, we would have to consider that in fact the arms reduced dramatically not once, but twice. Suddenly we have a proceratosaurid that survived many millions of years longer than we previously thought it did, and that it looked an awful lot like *Tyrannosaurus*. Such a dramatic shift is extremely unlikely (*Tarbosaurus* has far too much in common with other tyrannosaurines for this relationship to be overturned anytime soon), but this hypothetical example demonstrates the importance of understanding the relationships between species.

These and similar analyses are intrinsic to modern palaeontological studies because they allow us to track changes over time and space, so major contradictions like those of the Loewen phylogeny versus the 'standard' one are both intriguing and problematic. Only one model of tyrannosaur evolution can be correct (or, of course, possibly neither), and this is a rich vein of research for tyrannosaur experts.

For the sake of clarity, in the rest of this book I stick to the 'standard' model as the basis for anything discussing tyrannosaur evolution – though I do refer to Loewen phylogeny in key areas where it is relevant.

We now know what tyrannosaurs are and how to define them and their evolutionary history, but this group lived over a period of some 100 million years and across at least three and probably five continents. That's a lot of time and space that they covered, and quite how they were distributed, and which species or groups appeared when, is the next area discussed.

# CHAPTER FIVE

# Tyrants in Time and Space

With a good idea of what is and what is not a tyrannosaur, and the various lineages and how they are related to one another, it is possible to delve into what happened to these animals across the millions of years that they stalked the Earth. Later we examine the evolutionary patterns of changes in shape and size that occurred from the first appearance of the tyrannosaurs right up to the last of the group, but here we look at the pattern of their spread.

We do not know what the first tyrannosaur was. Phylogenies allow us to identify the species we think was closest to the split of the tyrannosaurs from the other theropods, but we cannot say if that animal was the original 'ancestral' species. Thus, even if we are lucky enough to ever find the genuine 'first', we won't actually know that we have done so – though of course evolution does not just split off a species as a ready-formed new lineage, but gradually separates out evolving populations, so a search for such an animal is rather futile.

What we *can* do, however, is to try and identify what that animal would have looked like, and when and where it would have appeared on Earth, based on what we know of the earliest tyrannosauroids and the basal coelurosaurs that they split from. The actual time of the animal's appearance might seem the hardest of these facts to pin down, but in some ways it is one of the simplest.

Contrary to popular belief, rocks and fossils are not ascribed an age based on radiocarbon dating or carbon isotopes. This is because the radioactive clock of carbon is only accurate to a few tens of thousands of years, and is thus of little use to anything older than about 50,000 years, and by extension, covers relatively few fossils, but instead is useful for ancient remains that have not yet mineralised and turned to stone. The principle, however, is not so far from that being used on much more ancient remains.

Many chemical elements exist in multiple forms, called isotopes; they have the same number of protons and electrons that give them their characteristic properties, but may have differing numbers of neutrons and so different mass. These various forms are given numbers that show their different atomic weights, so you will often see discussions of $^{238}U$ as a form of uranium, or $^{14}C$ as a form of carbon. The isotopes with extra neutrons are unstable and radioactive, and shed the excess baggage to revert to the more common stable form. Since this decay occurs at a uniform rate (the radioactive half-life), if we know the ratio of the radioactive isotope to the normal form at the start of a time period, and we know how much there is now, and the rate of decay, we can work out how far back the whole thing kicked off. In short, we can figure out roughly how old it is. Conveniently, some isotopes are stable over tremendously long periods of time, so we can date even very ancient rocks for those isotopes with long half-lives. Nowadays, these are measured with great fidelity and are calibrated against each other, and the best dates are drawn from multiple isotopes that decay at different rates so they can be compared. It's not uncommon to see rocks dated to 100 mya with an accuracy of a few tens of thousands of years. It's all very impressive.

Annoyingly, however, we can only date volcanic rocks, and few fossils lie within these (unsurprisingly). Therefore we tend to date the volcanic rocks that lie above and below the horizon that bears the fossil in question, and that gives a range within which the animal must have lived. In some cases this is quite narrow: where volcanic beds lie very close to the fossil, but in others they can be far above or below, leading to a wide range. Sometimes rocks haven't been dated well or have no dates assigned to them (we still have a general idea of when they were formed from cross-referencing to other rocks, but no exact figures), and that inevitably leads to still more variability. Geologists have also done a great deal of work correlating layers of rock between various locations, and palaeontologists have assisted using key fossils that are known to only occur in limited time periods, allowing them to align fossil beds all over the world. Thus if a new fossil-bearing location is suddenly discovered, it can be quite easy to work out which rock formation it comes from, or at least which it is closest to, and to get an estimate of the age.

A great many dinosaur-bearing rock formations have now been dated in some detail, so in most cases we have a good idea of the age of a given tyrannosaur, and in some cases we can discern the range of times in which a species existed.

## The oldest tyrants

The oldest tyrannosaurs are *Kileskus* and *Proceratosaurus*, both of which are about 167 million years old, but neither is known from very much material (from the Middle Jurassic Period). *Guanlong* from China,[15] however, is known from several near-complete specimens and gives us a much better idea of these early tyrants. It's just a little younger than the others, with the rocks it was discovered in having been dated to around 160 mya. We can also infer that the clade probably originated before this, even if we do not have any earlier tyrannosaur. Other theropods, like alvarezsaurs, which we know split from the main theropod line after tyrannosaurs, are known from the same beds as *Guanlong*, and even later theropods that are very close to birds are in rocks of the same ages as those that produced *Proceratosaurus*. If these various theropod lineages had already separated and had time to diversify by this point, it is likely that they had already been around for some time, and that the tyrannosaurs must thus have split off even earlier.

My personal best guess is that tyrannosaurs will ultimately turn out to have first appeared in the Early Jurassic, perhaps around 190 mya. This was not long after a major extinction at the end of the Triassic Period that rather shook things up on land, and ultimately probably helped the dinosaurs to go from one of many reptile groups to the dominant one on land. We often see major radiations after such extinctions, and this might well have facilitated some of the splitting of the theropods at this time, thus leading to the birth of the tyrannosaurs.

There are some fundamental biases of palaeontology that limit and affect what fossils we can find. The patterns of preservation are such that certain fossils are preferentially preserved (and indeed discovered), and others are heavily limited, so anything that we see is through this filter; this must be borne in mind when interpreting the often limited data that we have. The most obvious bias is that hard tissues (bones, teeth, shells and the like) are much more likely to fossilise than soft tissues (such as muscle, skin and internal organs), since the former don't tend to decay and the latter can break down quickly. So, for example, we have almost unlimited numbers of fossil ammonite shells, but in only a handful of cases is anything of the animal that lived in the shell preserved. Similarly, young animals tend to have less well-mineralised bones than adults, so these are not preserved as often as those of adults (though there are also bigger factors at play here, as discussed later). Small fossils are harder for us to find than large ones, so there can be biases against smaller finds, but very large fossils are unlikely to have been fully buried and are therefore most likely to be incomplete.

How and where an animal lived will also have a huge effect: for a corpse to become fossilised, it will need to be buried before it has fully decayed or been eaten by scavengers. This means that rainforests, despite their wealth of life, are terrible places for producing fossils because the rate of decay in them is so astronomical. Deserts, meanwhile, despite having low populations of few large animal species, can be highly productive in terms of fossils, giving the opposite picture of diversity and numbers of organisms to the reality. Similarly, any animals that inhabit coastal areas, or sites in and around floodplains and lakes, are much more likely to be buried and preserved than those that lived in upland regions. I do wonder if we are missing the dinosaurian equivalent of mountain goats, since mountains would probably produce few fossils. Basic biology and numbers also come into play here: there may be hundreds of herbivores for every carnivore, and small species tend to be more numerous than large ones. Additionally, we are limited by what rocks are available and accessible: dinosaurs are, for example, known from Antarctica and North Korea, but both are rather difficult to explore for different reasons. We also do need to access rocks that are exposed on the surface to be able to hunt for fossils, so any dinosaur-bearing rocks that are out to sea, or under soil or even towns, can't be explored (the dinosaur-bearing rocks of the Paris Basin are, for example, difficult to access in places for the obvious reason that there is a city on top of some of them).

Add to all this the fact that few individual animals are ever buried at all, or found by palaeontologists (in the field you typically find far more fragments of eroded and disintegrated fossils than decent bones, let alone whole skeletons), and it is something of a wonder that there are any dinosaur bones in museums, let alone entire halls of skeletons. The factor in our favour is, rather inevitably, time. Only one in a million animals may ever become a fossil, but a standing population of even a large-bodied species may easily number hundreds of thousands or millions of individuals, and each year they will produce millions of offspring, and over a few million years that the species might exist, the numbers of animals might well be in the billions or tens of billions, which gives us quite a bit to work with (assuming we can find them and dig them up).

The final relevant factor with regard to the quality of the fossil record is the age of the rocks. After an organism becomes a fossil, much might happen to it. Assuming that the rocks bearing the fossil are not buried under new layers of rock, or dragged to the bottom of an ocean by continental drift, they might be eroded. The longer they have been around the more likely this is to have happened, so the closer we get to the present, the better the fossil record tends to be. It's no surprise, then, that we have a better record of the Late Cretaceous than, say, the Middle Jurassic, and that the Mesozoic as a whole is worse for finds than more recent periods, but much better than the considerably older Cambrian (though of course individual fossil localities can be highly productive and very revealing).

All of this influences tyrannosaurs as much as any other clade, and in terms of finding early forms we incur almost every bias going: the first tyrants would have been small, few in number, carnivores, and living back in the Middle or even Early Jurassic. *Guanlong* is the closest we have to the origins of the group (at least in terms of a whole animal to work from), but given the raft of recent finds of dinosaur remains from the Middle Jurassic, there's a good chance that we will find more and better remains of animals like *Proceratosaurus*, which will reveal more about the particular part of the origin and early evolution of the tyrannosaurs.

The dating of our other tyrannosaurs reveals that they covered a wide spectrum of time periods. There is some bunching up of species because a number have been found together in productive areas (like *Alioramus* and *Tarbosaurus* in the Late Cretaceous of Mongolia, and *Dilong* and *Yutyrannus* in the Early Cretaceous of China), but overall there's a good spread of forms over the 100 million years of the tyrannosaurs' time on Earth. *Proceratosaurus* and *Kileskus* serve as both the oldest known tyrannosauroids, and are also the earliest of the procertosaurids, with the first tyrannosaurid not appearing until a considerable time later. Both *Albertosaurus* and *Gorgosaurus* date to around 70 mya for the albertosaurines, but the lineage must be older still because they split from the tyrannosaurines, and the earliest member of the tyrannosaurines is the recently named *Lythronax*, which is around 80 million years old. This does mean that these derived tyrannosaurids separated and radiated relatively close to the end of the tyrannosaur lines, with 'just' 15 million years to go.

At the very least, tyrannosaurs as a group were long lived and outlasted many other dinosaur lineages: from their first appearance they made it to the very end of the Cretaceous. *Tyrannosaurus* is literally one of the very last dinosaurs known, and its remains are found pretty much right up to the time of the great extinction event 65 mya. All those dramatic images that abound of a tyrannosaur in front of a burning world, or left as a charred skeleton on a burnt plain, are perfectly legitimate: the tyrannosaur line only ended with the mass death of so many of its relatives.

## Around the world in 80 million years (or so)

As well as existing for a good period of time, the tyrannosaurs also got around. While their fossils are most numerous and best known from North America, eastern Asia and, to a lesser extent, Europe, putative tyrannosaur remains have also turned up in Brazil and even in Australia. Given that palaeontological exploration of the southern continents is still in relative infancy, it is probable that there are more to be found – though the identity of these southern specimens as tyrannosaurs has been questioned and they are not unambiguously tyrannosaurian in nature. The specimen from Australia, for example, consists of little more than part of a pelvis, and while it has the enlarged pubic boot and other features that are classically tyrannosaurian,[16] with very little data it is somewhat ambiguous.

Of greater interest is the subject of which species appeared where in the world and when. By combining our knowledge of the ages of various fossils, their geographic location and the evolutionary relationships between them, we can trace the patterns of their spread around the world. Coupled with data on when the continents split from one another, it is possible to get an idea of how some groups may have evolved in relation to these grand events.

Relevant to this, it is important to try and establish if a given pattern of species across the continents is linked to either dispersal or vicariance. How did species get to be where we found them? Under a dispersal model, members of a lineage move to an already separated continent or land area, for example, the first animals arrived on the newly formed volcanic Galapagos Islands from the South American mainland. In the case of vicariance, species move with the land: when a continent or land mass separates off, it takes species that already live there with it, and their descendants remain there even if they later die off elsewhere. Vicariance is responsible for the presence of lemurs on Madagascar: they once lived on mainland Africa as well, and when they died off there only the isolated population on Madgascar was left.

Organisms can travel between continents, even if they are well separated, though of course this is rather easier for those that can travel long distances across water safely, such as birds and coconuts. In the Mesozoic, the continents were typically closer together than they are now, but even so there were various points where large land masses, such as India, were well separated from everything else, while other areas, such as Antarctica and South America, could be bridged with relative ease (Fig. 8).

Looking at the standard tyrannosaur phylogeny with the known locations of the various tyrannosaurs, the pattern is quite complex. Early forms are distributed across numerous continents, with the proceratosaurids and other early tyrannosauroids being found in Europe, North America and Asia. Starting with the proceratosaurids, as noted *Guanlong* dates from the Middle Jurassic and is from western China, *Procertosaurus* is from the same period (though slightly older) and from Britain, while *Stokesosaurus* is from the Late Jurassic of the US and is around 150 million years old. Thus even this first and early radiation of tyrannosaurs lasted for at least 10 million years, and the animals lived on three continents.

Fossils of various specimens and species of uncertain phylogenetic affinity complicate the picture, and items like the Australian pubis and the Brazilian material already mentioned suggest that tyrannosaurs may have spread far and wide at various times, right down into the southern continents. However, the main material still suggests that the group collectively was primarily based in the northern hemisphere, with animals like *Eotyrannus*, *Dryptosaurus* and *Xiongguanlong* appearing in Europe, North America and Asia respectively through into the Early Cretaceous. Moving up the phylogenetic tree a little, the albertosaurines are clearly based in North America, and appear to represent a localised evolutionary radiation that did not spread out of the continent; similarly, the alioramins are currently only known from east Asia. However, although the northern continents were spreading out ever further in the Early and Late Cretaceous, the tyrannosaurines appear in both North America and Asia.

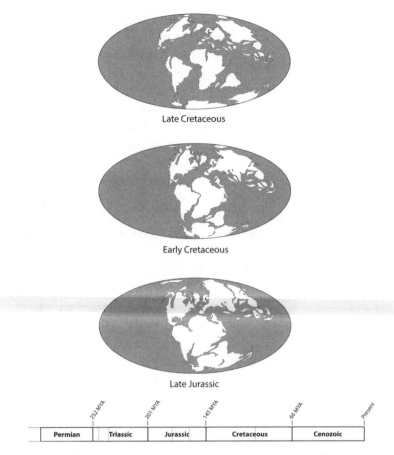

Fig. 8 The continents moved a fair bit over the 100 million years or so that the tyrannosaurs were around. Shown here are their approximate positions in (top to bottom and therefore going back further in time) the Late Cretaceous, Early Cretaceous and Late Jurassic. Below is a very simple timeline of the Mesozoic Period in which the dinosaurs dominated with the divisions within the Jurassic and Cretaceous marked.

Looking at the comparable dinosaur faunas in Asia and North America, we do unsurprisingly see a series of parallels: both contain ceratopsians, hadrosaurs, oviraptorosaurs, various dromaeosaurs, troodontids and alvarezsaurs, and one or two sauropods. However, overall each continent seems to have lineages all of its own. There are generally no close relatives of the North American animals in Asia and vice versa (though the hadrosaur *Saurolophus* is an obvious exception, as this genus is known on both continents), suggesting that generally not too many animals crossed from continent to continent. However, the tyrannosaurines must have done so at some point, and perhaps many times.

Referring to the tyrannosaur phylogeny in the previous chapter, we see that all the early tyrannosaurids (both albertosaurines and tyrannosaurines) are from North America, suggesting a possible origin for the group there, but later there is a burst of tyrannosaurine diversity in Asia (like *Alioramus*, though the origin of the alioramins might confuse this otherwise North American base). This means that at some point members of this group crossed into Asia and diversified, but right at the 'top' of the tree we have *Tyrannosaurus*, an animal that is very much American, so its ancestors must have got back across the Bering Strait and to the Americas.

Some of these patterns are complicated by considering alternate possible hypotheses for tyrannosaur phylogenies. The Loewen *et al.* tree mentioned earlier, for example, suggests alternate transitions at different times, but even so this would still result in multiple shifts for the tyrannosaurines across the Bering Strait. Clearly this group consisted of something approaching dinosaurian globetrotters (or at least, continent swappers), and it was moving around more than most other lineages.

One interesting subject is the faunal changeover of the Cretaceous. In later chapters we look at the changing fauna (including both competitors and possible prey species) that lived alongside the tyrannosaurs at various points in their history, but there is a key event in the Mesozoic that was thought to have prompted many of the changes we can detect. Sometime in the Jurassic flowers, an evolutionary novelty, were introduced to plants. Like many new lineages, flowering plants were initially small and rare, but starting around 125 mya (in the Early Cretaceous) we see their mass expansion, together with changes in many lineages of animals (including insect pollinators, for example). Classic images of giant sauropods eating tree ferns and cycads give way to hadrosaurs and ceratopsians browsing on conifers and other, more modern plants, while the birds expand, social insects appear, mammals diversify and other groups seem to generally get on with changing. This period is known as the Cretaceous Terrestrial Revolution, and it is thought that the mass changeover of plant life towards more modern flowering plants, called angiosperms, led to a sea change (except in the sea) in life on the land, with new lineages adapting and evolving to exploit them, and by extension others fading or going extinct as a result of them.

Much did change in the Cretaceous, and there were certainly major changes in dinosaurian fauna around the world. By the Late Cretaceous, for example, there were few sauropods left in the northern hemisphere (though they continued to do well in the south), while the ankylosaurs, ceratopsians and hadrosaurs made up a huge part of the herbivorous fauna, and the various feathered theropods called maniraptorans in all their forms really got going. However, it has recently been shown that the changes that appeared to have happened at this time were perhaps not as dramatic as previously thought: although the diversity of many of the newer groups increased in the Cretaceous, and came to dominate in the Late Cretaceous, they first appeared much earlier. Members of all of the maniraptorans were present in the Middle Jurassic at least (except birds, according to current records – although *Archaeopteryx* is known from the Late Jurassic), and other major lineages, such as ceratopsians, were also around this early. It is probable that the revolution did much to prompt the changes in this fauna, but the idea that it sparked new dinosaur diversity (or indeed that new dinosaurs sparked the rise of flowers, as was once suggested) now seems unlikely. It is likely that a more steady progression occurred.[17]

Certainly, the tyrannosaurs covered a large area for much of their existence, and while they never truly became established down south, in the Cretaceous generally and the Late Cretaceous in particular, they were numerous in terms of species and individuals. Despite some gaps in our knowledge, the overall pattern of dispersal is quite clear, and while new discoveries may provide more information, it seems unlikely that we will turn up some previously unknown radiation in North Africa, say, or that there were major interchanges between South America and Asia. Nevertheless, it seems that the tyrannosaurs got around: although other groups also ended up with similarly wide distributions, we don't currently see the kind of back-and-forth shuttling in them that we do in the great tyrannosaurines.

Having described the tyrannosaurs in their evolutionary context, in the next section we delve into the anatomy of these animals in more depth. Tyrannosaur research is primarily based on the bones themselves, and by looking at these we can learn about soft tissues such as the muscles, lungs, eyes and brain, and calculate the animals' growth, how fast they could move, how they ate and much more.

# PART TWO

# MORPHOLOGY

# CHAPTER SIX

# Skull

The average reptile skull contains numerous complex and often fragile bones, so finding a good one is a rarity. However, it is the most important part of almost every fossil skeleton: it contains a disproportionate amount of anatomical information because it is so key to many aspects of an organism's biology. The head houses the major sensory systems, the brain to process their signals and send out instructions to the rest of the body, and the main feeding apparatus. Understandably then, palaeontologists can be very pleased to get even a partial skull of an animal, and one in good condition can produce huge amounts of data. This is perhaps particularly the case for the tyrannosaurs,[1] given the extent to which the skull was modified from the ancestral form in the later groups, and the emphasis on the head for finding and consuming food due to the reduction of the arms.

The overall shape of a skull can give some immediate clues to the lifestyle and ancestry of an animal, and the tyrannosaur skull is obviously that of a carnivorous archosaur (Fig. 9). The presence of the large antorbital fenestrae in the sides of the face between the orbit and naris mark out these animals as archosaurs, while the large and serrated teeth scream out 'carnivore'. More than that, however, the orbits are set in such a way that the eyes point forwards, giving the animals binocular vision and making them excellent at judging distances. We see this feature in some arboreal animals (including humans and other primates), but it is found primarily in carnivores. Herbivores have laterally positioned eyes that give them something close to wrap-around vision and enable them to look out for threats, but predators have forwards-facing eyes for measuring the distances to their potential meals and calculating strikes. This characteristic is further exaggerated in the tyrannosaurids, and especially the

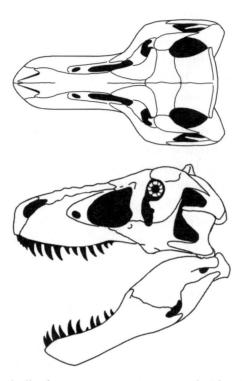

*Fig. 9 The skull of a tyrannosaur in top and side views. This is* Tyrannosaurus *which has more of a T-shaped profile when seen from above than do other tyrannosaurs and helps the eyes face forwards.*

tyrannosaurines, in which the back of the skull is expanded to give something close to a T-shape when seen from above, with a long and relatively thin snout, but one that also modifies the orientation of the eyes still further. The snout tip is rather pointed in early tyrannosaurs, but more rounded upfront in the tyrannosaurines.

In the case of the derived tyrannosaurs, we can also see how they appear overbuilt as a whole: there is a lot of bone there, and they are not just composed of thin struts like many other theropods. Although there are numerous large openings in the skull, the unit as a whole is robust. At the most extreme, animals like *Tyrannosaurus* had among the largest and strongest skulls of any terrestrial animal. The fused nasals that characterise tyrannosaurs are symptomatic of this overall construction, strengthening the skull and allowing for a powerful bite. Indeed, derived tyrannosaurs had extraordinarily powerful bites, as can be seen from the bite marks they inflicted on bones when feeding. Mechanical analyses of their skulls and teeth also reveal how strong these were, and in particular how resistant they were to the kinds of force induced by strong bites.[2] The joints between the various skull elements could move a little, which further helped to dissipate the forces of heavy biting.

The nostrils were laterally positioned on the skull, but there may not have been a septum (cartilage wall separating the nostrils) in between them – it is not impossible that a perfect side-on view would have allowed you to see through this part of the head. However, this would have reduced a tyrannosaur's ability to accurately trace scents, so tyrannosaurs probably did possess a septum. The lateral position of the nostrils, in combination with a septum, would have enabled each nostril to gather a slightly different part of the air, and helped give directionality to the sense of smell: if an odour is stronger on one side than the other, you know which side it is coming from, and tyrannosaurs certainly had an excellent sense of smell. Internally, the nares lead into the main part of the inside of the skull (between the maxillae and above the palate), and this would have been part of the sinus cavities – for all that bone, the skull was largely hollow and not as heavy as you might expect. Some of the bones of the skull were pneumatic, too, with extensions of the pneumatic system (sinus, trachea, lungs, air sacs) extending into them.

Almost all tyrannosaurs show some combination of rugosity (a rough and bumpy surface) to the nasals and surrounding bones, and little horns that sprout up over the eyes. These would have been covered with a layer of thickened keratin (the same material that makes up hair, feathers, scales and fingeranils or claws), which would have increased their size and made them somewhat tougher. This may have been to exaggerate them as ornaments, or to offer some additional protection for the head; as discussed later, tyrannosaurids took quite a lot of punishment to the face. Exceptionally, the proceratosaurids had an exaggerated crest of bone in the midline of the skull. This was rather fragile, and also pneumatic, and was made up of thin bony supports, but again it might well have had quite a bit of keratin over the top.

The hornlets were not large but did protrude slightly forwards and out to the sides, which would have given the eyes a rather hooded appearance, but the actual eyeballs would not have been sunken into the face as is often shown in illustrations. The orbital cavity is large in tyrannosaurs, and especially so in the tyrannosaurines; *Tyrannosaurus* and its nearest relatives would have had among the largest eyes of any known terrestrial animal.[3] They appear small because of the sheer size of the head, but absolute size is more important when it comes to the eyes than proportional size, and the eyes of the big tyrannosaurs were very large.

Most vertebrates have a ring of tiny bony plates that sit over the eyeball and help support it, and tyrannosaurs were no exception (mammals are actually rather unusual in lacking this feature), but these little pieces were so fragile that they are rarely preserved. Although they appear in illustrations in this book, they are not normally recreated on mounted fossils. When they are shown, however, they help give a still better idea of the size of the eyeball overall, and the hole in the middle corresponds closely to the iris. At the back of the skull there is a slight indentation halfway up each side in which the external ear would sit; it would not be located in any part of the hole that sits below and behind the eye on the side of the skull, as is often incorrectly assumed.

## Brain boxes

Moving inside the skull, the brain would be entirely encased in a series of bones, collectively termed the braincase, which sits roughly between the eyes and is almost suspended within the cranium. As this fits the brain so closely, the internal size and shape of the braincase effectively produces a mould of the brain itself. A cast taken from it (or nowadays a 3D scan of it) can reveal a number of important details, and we know what the brain of *Tyrannosaurus* was like from a scan of the skull of 'Sue'.[4] Such a scan can indicate the structure of the inner ear, and thus even point to the kinds of noise (high or low pitched) the ears were attuned to. Sadly, this information is not currently known for tyrannosaurs, but the potential for it to be discovered exists. I would expect tyrannosaurs to have a good range overall — large tyrannosaurs would be interested in the high-pitched sounds of small prey and the low-pitched sounds of other tyrannosaurs, while small tyrannosaurs might focus on higher sounds generally, low pitches might indicate a threat and they will keep an ear open to them.

The individual parts of the brain appear in a strict arrange-ment in animals, so it's pretty easy to work out which bit is which, and if certain areas are especially enlarged. *Tyrannosaurus* famously has an enlarged olfactory bulb, so would have had an especially good sense of smell, and this was likely true of other tyrannosaurs too. Various openings pierced the braincase to allow different nerves and blood vessels to enter, and there was a huge opening at the back for the main nerve cord to extend from the brain to the rest of the body. Above this, a huge ball of bone on a stalk — the occipital condyle — was the articulation point of the skull with the rest of the body, and it is thus a major feature that links the body together.

A common and very simplified measure of intelligence pitches brain size against body size. A bigger brain is required to run a bigger body, so larger animals have larger brains than smaller ones, and any brain that is bigger than would be expected for an animal of a given size might point to a smarter animal (humans, for example, have disproportionately large brains). Tyrannosaurs are no exception to the general trend in animals, so were unlikely to have been exceptionally bright, but that doesn't mean they were stupid. After all, a great many animals with similarly sized brains can perform some difficult tasks, and engage in complex and impressive behaviours such as problem solving.

Moving down to the mandible (the combined bones that make up the lower jaws), this was relatively thin in early tyrannosaurs, but like the skull it became more robust and deeper in later forms. Again, this would have increased the ability of this part to resist heavy forces when biting. The dentaries can be slightly concave along the upper edge, and may match the slightly convex curvature of the maxillae, which, like curved scissors, would probably help to even out the delivery of forces during use. The two sides of the mandible are often separated in tyrannosaur fossils, and they are not bound together with the anterior parts of the dentaries fused together as in many animals (including humans), nor are the dentaries fused to the bones at the back of the mandible. This has led to the occasional suggestion that the two sides of the mandible could move apart a little, and that the dentaries could flex outwards relative to the rest of the jaw bones. This would have helped to increase the size of the gape and thus the ingestion of larger bites of food. However, in fact these various bones would have been held together pretty firmly by a series of ligaments at the tips of the dentaries (scars for these ligaments are clearly visible on well-preserved bones) and others further back. While there would probably have been a little wiggle room in there, which might have helped to dissipate the forces of a heavy bite, they did not move around separately and change the shape of the jaw.

Ligaments that bind bones to other bones, and tendons that anchor muscles to bones, leave traces of their locations on the bones to which they are attached. Combined with our knowledge of the patterns of muscles in living archosaurs, it is possible to work out where major muscle groups attached and what their primary functions were. Those of the skull are rather obvious – some major muscle groups at the back of the skull closed the jaws and delivered bites, while a much weaker set of muscles opened the jaws. Various other muscles at the back connected to the neck and allowed the head to change in orientation.

Several sets of muscles connected the cranium to the mandible to open and close the jaws. Those connecting the back of the skull to the posterior part of the mandible acted as depressors, opening the jaws. Muscles can only contract to make themselves shorter (to get longer they must relax, then let another set of muscles do the job to return them to their original state). Thus the connection at the very back of the jaw, and behind the articulation point for it with the rest of the skull, allows it to be pulled open via a contraction. Since this doesn't require much effort (the jaw is also being pulled down by gravity, and you don't need to open the jaw too hard or against anything), it needs proportionally little muscle. This is why, not that I'd advise trying, you can hold the jaws shut of animals like crocodiles and lions with only one hand, since the power is in the closing and not the opening of the mouth.

The muscles that close the jaws sit in front of the articulation point and are much more numerous and larger than those that open the jaw; they may connect to the top of the skull and to the palate. This gives the real power to the jaw closure, and the later tyrannosaurs, with their literally bone-crushing bite, obviously possessed a lot of muscle there. Unlike in mammals, however, the muscles were largely held inside the skull and attached inside it. Thus when you clench your teeth, the muscles can bulge out in your cheeks and it is rather obvious what is happening, but this would not have occurred in dinosaurs – though there might have been a bit of bulge in places. This feature in part gives the dinosaurian head its different appearance from the heads of mammals.

One free-floating pair of bones in a tyrannosaur skull would have been the hyoids, which don't attach to any other bone in the body. These have yet to be recovered for any tyrannosaur as far as I know. Typically, a pair of hyoid bones is fused together into something of a Y shape, and when present (they are rarely preserved in fossil taxa) the bones would be located between the dentaries. In a living animal, the hyoids act as supports for the tongue but they can be thin and fragile bones which may explain why they have yet to be found.

The jaw-closing muscles cannot be situated at the front of the head because they would get in the way of the eyes and nose, and indeed make the opening of the mouth too small, so they are located at the back. However, as a result of basic lever principles, the further away from this point you travel, the less force is delivered. Thus tyrannosaurs with proportionally long heads, such as *Alioramus*, would not have had the same bite power as an animal with the same amount of muscle, but a shorter head. This also means that the most powerful bite would have been delivered closer to the back of the jaws, though this must be traded off against the fact that the jaws only opened so wide, and it would have been hard to get anything that far back into the mouth. In short, the biggest bites would have been delivered somewhere in the middle part of the maxilla, where the trade-off between power and practicality would meet. It's therefore no surprise that the largest teeth in the tyrannosaur jaw are those of the front third of the maxilla on the upper side, and in the middle of the dentary in the lower jaw.

## Quite a smile

Tyrannosaurs are unusual for theropods in having different types of teeth in their jaws. While the teeth of the premaxilla in most theropods are little different from the other teeth, in the tyrannosaurs there is a marked difference between these teeth and those of the maxillae and dentaries. There is a total of eight premaxillary teeth (four in each side), and these are sometimes described as being 'incisiform' – that is, they are shaped like the incisors in mammals. This shape makes them very resistant to bending forces from multiple directions, and they would have been strong. As discussed later, this would also have made them well suited to their use in feeding, for scraping across bones to remove the meat.

The numbers of teeth in each maxilla and dentary vary in the tyrannosaurs, with each having between 11 and 18 teeth, and there typically being more in smaller, older tyrannosaurs, and fewer in the more recent and larger forms. The first teeth of the dentaries are typically small and rather like the premaxillary teeth in shape, but the rest of the dentary teeth and those of the maxillae are rather more 'normal' for theropods. Actually, the ancestral archosaur tooth is quite similar to a normal theropod tooth, and its shape and pattern are common in many carnivorous lineages (including various extinct crocodilian lineages). The teeth are relatively flat and curved back, often with the curve at the front of the tooth being rather stronger than that at the back, which is straighter. In the case of tyrannosaurs, these teeth are typically thicker than those of many other theropods. Animals like *Allosaurus* and *Velociraptor* have teeth bordering on a blade-like pattern. They are quite thin, with an allosaur tooth 10 centimetres long measuring perhaps a centimetre at the thickest point. In the tyrannosaurs the teeth are much thicker – more than double that in some cases – making them strong and resistant to breaking when going in for those big bites. The teeth in a tyrannosaur jaw don't change only in shape and form, but also in size. The premaxillary teeth are fairly uniform, but those of the dentary and maxillary start quite large, then there are several especially large teeth, before the teeth get progressively smaller along the tooth row, with the last teeth being rather small.

An oval cross-section of most teeth reveals that they functioned rather more like spikes than knives, and would have delivered puncture wounds rather than cuts. However, that is not to say that these teeth could not slice as well. All tyrannosaur teeth (indeed most theropod teeth) have edges lined with serrations to help them cut. These serrations are called denticles, and together they form a line on the tooth termed the carina. The denticles are rather small, and in appearance they look a lot like one side of a very small zip. Comparisons to saw blades and steak knives are accurate because they largely operated as such, helping to slice open skin and muscle tissues. Indeed, experiments with tyrannosaur teeth prove this – the denticles sever individual muscle fibres, helping to ease the passage of the tooth into the meat, and to deliver a deep and effective bite.[5] However, the serrations themselves are on a much smaller scale than those on a steak knife: each was under half a millimetre in length.

The denticles are slightly proud of the tooth surface to help produce that edge, though the slight lateral compression of the maxillary and dentary teeth already helps to produce a front and back edge for the carinae. However, while the carina of each tooth runs down the front edge, the posterior carina can take some interesting detours in the tyrannosaurs. In many animals the posterior carina of the maxillary and dentary teeth has something of a twist to it, so as we move from the tip of the tooth towards the base it turns away from the tooth edge and moves to the lateral side of the tooth. In other animals the rear carinae split into two, with each bit diverging on its way down the tooth. Quite what advantages these features may bring is not known; they could just be random features that offer neither an evolutionary advantage nor a disadvantage, and so persist. The fact that at times they appear almost at random in some individuals of various species would suggest that there is no great selection pressure on them to expand or to be lost.

The roots of tyrannosaur teeth were large, with each being around twice the length of the tooth crown. This typically anchored each tooth firmly within the jaws, though tyrannosaurs, like other archosaurs, cycled through teeth fairly regularly. After some use the teeth would become worn, and the sharper tips and carinae would wear away. As a tooth aged the root would be partially reabsorbed, leaving the crown loose in the jaw and held in by just a few ligaments. A new tooth would form underneath and push up, eventually forcing out the old tooth and allowing the new one to emerge. A tyrannosaur could therefore consistently have relatively new teeth in the jaws at all times. Old teeth would be dislodged during feeding or other activities, and this probably explains at least partly why shed theropod teeth may be found alongside bones that show feeding marks. It also means that on any given day, it's unlikely that a tyrannosaur had every tooth sitting perfectly in its mouth; mounted skeletons in museums, and images of tyrannosaurs and other theropods in art and pop culture, inevitably display a perfect and pristine line of teeth, when in fact in a living animal there would almost certainly have been a few missing teeth, as well as newly emerging teeth poking through and not in the final position. A single tyrannosaur that lived for a decade or more might thus go through hundreds, or even a thousand or so, teeth in a lifetime; theropod teeth can be very common, even in places where the number of individual animals is low.

Skulls are complex structures, combining numerous bones, teeth and a wealth of soft tissues (brain and nerves, eyes, blood vessels, muscle, cartilage and more). They are key in systematic and taxonomic work, as well as providing essential information on the behaviour of a skull's owner. The wealth of tyrannosaur skull material – few taxa do not have at least some good skull parts – makes skulls rich sources of data, and provides a comprehensive understanding of tyrannosaurs' biology. Heads are, however, only part of the deal, and from here we move down to the main body.

# CHAPTER SEVEN

# Body

The body of a tyrannosaur comprises the majority of the animal in terms of both length and mass. The spinal column runs as a continuous series of vertebrae from the very back of the head right the way through to the tip of the tail, and it provides the main support for the body and an anchor point for the limbs, as well as holding up the head. The body also contains the majority of the major organs, which are discussed briefly later. Since the head bone (described in the previous section) is connected to the neck bone, we begin the coverage of the body with the first of the cervical elements.

The first of the series of cervical vertebrae is the atlas, and as the name suggests it supports the skull. It is a highly modified and unusual vertebra that looks quite unlike any other. Behind it is the axis, another oddity and one that allows a lot of the free movement of the head in relation to the rest of the neck. Most tyrannosaurs had rather long necks, but as they increased in size, and in particular as the skulls got larger and heavier, the neck shortened in the tyrannosaurines, thus reducing the distance of the head from the legs and preventing the animals from being too front heavy and tipping forwards. All tyrannosaurs had 10 neck vertebrae, including the atlas and axis. The resting posture of the neck would be curved into something of an S shape, rising up from the body then arching over, with the head held pointing slightly down.

Aside from the atlas and axis, the other neck bones are fairly typical for vertebrates. The lower part of each is a rather squat cylinder called the centrum, and this provides the connection and articulation between each vertebra in the series. On top is a complex of bony projections collectively called the neural arch. This provides anchorage for muscles that support the bones and allow them to move, as well as some extra articulations that extend forwards to connect with the neural arch of the vertebra in front to give them stability. Between the centrum and the neural arch is a hole through which the spinal cord runs, connecting the brain to various parts of the body. Between each centrum is a pad of cartilage that provides some cushioning when the bones move, ensuring that they don't grind together. Each pad is rather disc shaped – these are the discs that slip and cause humans pain if they prevent the bones from moving properly, or worse put pressure on the spinal cord. The spine thus provides support for the body, as well as a means of articulation to provide flexion and allow movement, and a strong protective case for the spinal cord.

There are also some short cervical ribs that articulate with the cervicals and give some support to additional muscle groups. Both the muscles and various tendons help to hold the neck together and provide support for it, and well as allowing it to move. Actually, the tyrannosaur neck was not that dissimilar from that of predatory birds – the bones were anchored in a similar way and the kinds of movements they carried out were similar, especially when moving the head up and back during feeding.[6]

After the cervicals come the 13 dorsals. They can generally be identified by a slight alteration in the shape of the neural arch, and a change in position of the articulation of the ribs. These are the 'true' ribs and much larger than the cervical ribs; they are gently curved along their length and help to delineate the outline of the chest cavity. The dorsal vertebrae are more strongly bound together than are the cervical vertebrae, permitting less flexion and thus providing greater support for the body as a whole.

### Belly bones

Along the underside of the chest lies an odd set of bones called the gastralia. These are largely free floating in the body (that is, they are not directly articulated with other bones), and provide some support for the ventral part of the chest. In most dinosaurs they lie in a series perpendicular to the spine and comprise very thin rods of bone; even in the giant sauropods they are not much thicker than a few strands of uncooked spaghetti, and they are rarely preserved or recovered. In later tyrannosaurs, however, the gastralia are odd in size, shape and articulation compared with those of other dinosaurs. Each is almost like a rib in thickness, and the whole piece is roughly triangular in shape, with the thickened end being in the middle of the body. Rather than lying in a series of parallel lines, the gastralia overlap each other on alternate sides, producing a long series of interlocking Vs when seen from below; some may even fuse together, making them still more solid. It's a most unusual arrangement, but how and why it evolved is unknown. It would presumably have helped to support the huge and heavy chest, especially when the animal was lying down, though other equally large and heavy theropods never evolved anything like this, and why tyrannosaurs should be so different has yet to be investigated.

Returning to the spinal column, after the dorsal vertebrae are the five bones of the sacrum, which consists of a block of fused vertebrae. Many animals have surprisingly small sacra, including large ones such as elephants, but in the dinosaurs there are typically five or more vertebrae and they are quite large. In young animals the various sacral vertebrae are separate from one another, and they look rather like dorsals but that are positioned between the hips. However, as an animal grows, the centra all fuse to each other (often the neural arches fuse to each other as well), and become one solid block of bone. Lateral outgrowths of the bones, called sacral ribs, also fuse to each other, as well as to the respective sides of the ilia. Thus in adults, the pelvis is collectively made up of one massive, robust piece, and it is generally the largest single unit of any vertebrate skeleton. The sacrum is relevant to the transmission of forces. The weight of a heavy biped like an adult tyrannosaurid is taken entirely on the legs (and when walking or running on a single leg), and is transmitted through the femur. Were the ilia not firmly bolted to the sacrum, the weight of a dinosaur would surely push them up through the back of the animal, with the torso slumping to the floor.

Continuing down the line, after the sacrum come the caudal vertebrae of the tail. Dinosaurs are often reconstructed with very long, lizard-like tails, but there is actually huge variety in tail lengths, in terms of both the number of vertebrae and the total length of the tail. Some dinosaurs clearly had huge tails: the giant sauropod *Apatosaurus* had at least 82 caudals, including some huge ones, whereas the oviraptorosaur *Caudipteryx* had 22, many of which were quite short. Due to a combination of factors (related to how many caudals there were, their size and how firmly they were held together), even partly complete tails for dinosaurs are very rare, and truly complete ones are almost unknown. Fortunately, one of these is a specimen of *Gorgosaurus* that has every single bone of the tail, with 37 caudals.[7] However, it is unwise to extrapolate this even to other specimens of *Gorgosaurus*, let alone other tyrannosaurs. Despite the paucity of data on complete dinosaur tails, what we do know is that both the number of vertebrae and the length of the tail varied considerably, even within species. In a tyrannosaur, the first few centra of the tail are rather short, then there are some relatively long ones, then they taper off, getting shorter and shorter down the tail towards the tip. Around halfway down the tail, the vertebrae begin to get very simple and the complexity of the neural arch reduces enormously, until eventually there is little more than the centrum left and the very last vertebra in the tail is little more than a hemispherical nub. Since there is progressively less bone, muscle and blood vessels moving down the tail, less material is needed to hold it together and keep it stiff, so all of the articulations fade away as they provide little more than unnecessary weight.

Below the tail vertebrae hangs a series of smaller bones called the chevrons (or the haemal arches). These each have a pair of articulations that connects to each vertebra above, though as a result of their shape they sit more between the joints of the vertebrae than directly below a single centrum; then they each have a longer piece of bone that extends posteriorly, so that they look rather like boomerangs or L shapes in lateral view. Between the centrum and the chevrons lie some of the main arteries that supply the tail and the muscles that connect to the hindlimbs, so in effect the chevrons act like a neural arch on the underside, but they protect the blood supply rather than the nerve cord. The shapes of these various chevrons change down the length of the tail: they get shorter and more simple in shape towards the end and eventually stop altogether. The first few vertebrae of the tail also lack chevrons, as this is where the cloaca (the urogenital opening) would have sat.

The joints between the tail vertebrae are less restrictive than those of the dorsal vertebrae, and the tail would have had some flexibility. It would not have been whipped around freely like that of a cat, but would have had a fair bit of lateral movement and curve to it when necessary. Overall, though, it was rather stiff and would have been held horizontally according to its articulations and muscles. The old-style images of tail-dragging dinosaurs have now been firmly disproved for two reasons, firstly because – as discussed later – we are confident that they had a more upright posture than was initially imagined, and secondly because the supporting articulations of the tail and the muscles would have helped to keep it relatively stiff. Additionally, fossil dinosaur-track sites only very rarely show drag marks for tails (and then only for the tip of the tail), so the idea that half of the tail dragged on the ground is clearly incorrect.

This makes a lot of sense as the tail provides two key locomotory activities, and a dragged position would rather inhibit these. First of all, in the bipeds it provides a counterbalance. If the body is to be held largely parallel to the ground and with the feet below the hips, then it will tip forwards without something behind it to hold it up: the tail provides this counterweight. The tail might not look big enough to do this job, but as we will see, the body is not as heavy as it appears, and as the tail is typically rather longer than the head and body, some mechanical advantage is gained as part of it is further away from the feet than is the head. Secondly, the tail provides the anchor point for the huge caudofemoralis muscles that run to the base of the femur. This is the main retractor muscle group that pulls the leg back, and thus moves the animal forwards. In short, it was the key muscle that enabled a dinosaur to walk or run, and wouldn't have functioned effectively if the first part of the tail had been near parallel to the femur with a tail dragging on the ground and the dorsal series of vertebrae raised up. It has recently been shown that this muscle was larger than previously thought (Fig. 10) and would have produced something of a bulge at the base of the tail where it was anchored.[8]

There are various other muscle groups that run through the neck, body and tail to provide support and movement to an animal, and these are supplied by a network of blood vessels, nerves and various other tissues. However, here I want to turn to the main internal organs of the tyrannosaurs. Much of this information is based on little more than reasoned speculation (essentially all vertebrates have some form of kidneys, and it is hard to see how tyrannosaurs could function without them, so it is safe to assume they had them), or informed extrapolation (holes for blood vessels in bones show where they went and indicate what they were probably connected to), but in some cases various exceptional fossils help us out a great deal.

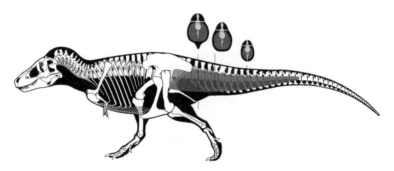

*Fig. 10 The huge and powerful caudofemoralis muscle that attached midway up the femur and spreads out across much of the tail providing retraction to pull the leg back and push the animal forwards. Several cross sections are also shown.*

## Squishy bits

A few years ago there was great excitement about the discovery of a fossilised heart of a dinosaur. Such a thing is not impossible – internal organs can be preserved on occasion if little enough decay has occurred and enough mud or other sediment coated the organ in question, or a natural mould might form so that even if the organs rot and are not fossilised themselves, their original shape is fossilised (rather like a footprint can be). Sadly, it ultimately turned out to be just an odd nodule of rock, but its size, shape, position in a skeleton and general appearance did make it a viable candidate. Both living crocodilians and birds have four-chambered hearts that are somewhat different from those of other reptiles, suggesting that dinosaurs had something similar. There are some differences between bird and croc hearts, however, so it is hard to know if dinosaurs were generally more croc-like or bird-like.

Much better understood are the lungs and air-sac system. As previously mentioned, theropods and sauropods were generally pretty pneumatic. In addition to the paired lungs and sinuses that almost all terrestrial tetrapods have (a few amphibians have effectively lost their lungs, and snakes have only one), these animals had extensions of the lungs that ran not just through the body cavity, but into many of their bones. Holes in a bone would pierce the outer surface, then expand and take up most of the internal volume of the bone. These pneumatic dinosaurs (and both modern birds and the extinct pterosaurs) had huge cavities in some of their bones; even the huge cervical vertebrae of some of the big sauropods could be more than 90 per cent air by volume and just a few per cent bone. Being hollow did not make them weak, however, because huge numbers of tiny bony struts called trabeculae would provide strength – a little like the spokes strengthen a bike wheel – and a bone section might look like a honeycomb. Many long bones (like the femur) are hollow in animals: if you cut open a cow femur you will see that the middle is hollow, but the cavity is both rather small compared to the pneumatic cavity of a dinosaur, and normally full of marrow, and it certainly doesn't turn up in bones like the dorsal vertebrae.

The air sacs pervaded many dinosaur bones, and in some cases almost all of a skeleton has some kind of pneumatic extension in it. In the case of tyrannosaurs it was concentrated in the vertebral column and associated bones. *Tyrannosaurus*, for example, had pneumatic openings into the cervical vertebrae and cervical ribs, the dorsal vertebrae and ribs, the sacral vertebrae and the first few caudals (though here the openings were rather small). Over time in various groups, these sacs were pushed further and further down the body, and some non-avian theropods even had pneumatic parts of shoulders and arms, pelvis and legs, quite some distance from their origins in the chest. There are multiple groups of air sacs and each set pervades only certain bones, so we can track their development and extension through the skeleton.

Importantly, in dinosaurs the passage of air through the lungs and air sacs was rather different than it is in mammals. We inhale to inflate the lungs, and the air fills them. Our lungs have numerous branches and sub-branches, and eventually terminate in huge numbers of little spherical structures. Here gas exchange occurs, oxygen enters the bloodstream and carbon dioxide leaves it, then we exhale and push the air back out the way it came in. This is bidirectional airflow: the same 'breath' of air travels both in and out of the same passages in the lungs.

In the birds, however, these various sub-branches in the lungs all connect to one another and thus are less a kind of giant set of tiny balloons, and more like a maze of tiny tubes. The lungs are also connected to the air sacs, and collectively this provides an alternate pathway. When breathing in, air is drawn into the posterior air sacs, and from there it goes into the lungs, then into the anterior air sacs and ultimately out again. Thus the flow in the lungs is unidirectional: the air travels only one way though most of the system, but the air sacs are really only there to help the flow, and gas exchange only occurs in the lungs. Many of the air sacs are therefore dead space; while some in the body cavity are used in breathing, those branches into bones are just there and won't normally see much, if any, airflow.

Dramatically, recent research has shown that alligators and some large lizards can breathe like birds, despite lacking air sacs.[9] Divisions of the lungs somehow allow them to act as both lung and air-sac groups in birds, and thus airflow is essentially unidirectional. This provides excellent support for unidirectional air flow as a breathing type in dinosaurs, and suggests that it evolved before the origin of the archosaurs and was thus at least originally universal in dinosaurs. The presence of numerous air sacs in theropods suggests that this characteristic was retained and even expanded upon: the avian lung is thus also a dinosaurian one.

Little is known about the digestive system of dinosaurs, but there is at least a bit of data for theropods. The position of the stomach can be roughly determined by the position of masses of consumed bones in the upper part of the chest cavity of various theropods. We know a little about the shape of the intestines thanks to an incredible specimen from Italy of a tiny theropod called *Scipionyx*. This is a compsognathid (and thus a close relative of the tyrannosaurs), and it was preserved with various organs, or at least impressions of them, including most of the digestive system.[10] It was overall rather short, as would be expected: both *Scipionyx* and the tyrannosaurs were carnivores, so they would have had relatively short and simple digestive tracts, since meat is generally quite easy to digest compared with plant matter. The fact that even *Tyrannosaurus* coprolites (fossilised faeces) still contain identifiable fragments of bone tells us that they did not have super-acidic systems as do some animals, so they were perhaps ordinary rather than exceptional when it came to how they broke down their food.

One last major set of organs relates to excretion and reproduction. As noted above, tyrannosaurs (like all reptiles and birds, and indeed some mammals, such as the platypus) have a single cloaca. This is a combined exit for the genito-urinary tract and the digestive system. In short, everything that is going to come out, comes out of this. Its location would naturally be at the base of the tail and above the posteriormost part of the ischia. We know dinosaurs excreted faeces because we find their coprolites, but they may have produced something like that of birds and crocodiles, in which the faeces and urine are mixed (hence the white streaks in bird droppings, because they are full of urea), or they might have urinated separately as various birds do, at least on occasion.

The cloaca would also have been the exit for eggs in females. Eggs would have been produced in two ovaries and developed in two oviducts, unlike in birds, which limit themselves to one. We know this in part because many maniraptoran theropods seem to have laid eggs in pairs in the nest, and a troodontid is known with a pair of eggs in the body, showing that the two eggs developed at once. In the tyrannosaurs multiple eggs probably developed simultaneously (as seems to be common in many therapod dinosaurs), but notably this would have happened in both oviducts simultaneously.

A male tyrannosaur would have to get its sperm over to a female and into her cloaca. This may seem impossible given the general structure of tyrannosaurs, but plenty of reptiles and birds mate with little more action than pushing their cloacas together. However, the size of many dinosaurs (and indeed their shape when it comes to things like the armoured stegosaurs) would have made this awkward, and something to help bridge the gap may have been employed.

Actually, a number of animals with cloacas have produced some form of 'intromittent organ'. This term is not some unnecessary piece of prudery, but to actually distinguish this from the different mammalian penis. The organs in birds and reptiles that have them are analogous (they have the same function), but not homologous (they do not share a single evolutionary origin) with the penis, so should not have the same name. Intromittent organs may be short and simple, or quite long and complex structures, depending on quite how they are used and the mating issues at hand. For large tyrannosaurs, something might well have been employed to help a male reach across to a female, but we are limited to trying to see if we can get the skeletons to fit together to work it out (a pair of tyrannosaur skeletons was once mounted in a mating posture for a museum display). For an insight on just how odd an intromittent organ can get, I highly recommend looking into the sexual biology of ducks. An explosively inflating organ that is both longer than the animal that bears it and helical in shape is really only the start. Ducks are, well, 'interesting' when it comes to sex.

Key to getting those animals together is the hips and legs, and how well a multi-tonne tyrannosaurine might be able to work with its mate to bring the two important parts together. Balance would be one issue, but flexibility and the overall use of the limbs (both fore and hind) would be key. This subject, then, provides a convenient transition to the next section, on the final major bony part of the body: the limbs.

# CHAPTER EIGHT

# Limbs

Finally we get to tyrannosaur appendages, namely the arms and legs, more formally called the forelimbs and hindlimbs. In tyrannosaurs the forelimbs are not used for walking, so the colloquial use of 'arms' for forelimbs works well (given that the term 'legs' is often used to indicate than an animal walks with them). Limbs need something to work off in order to move, so they are connected to the main structure and support unit of the body – the spine. Each limb is articulated with a series of bones that attaches it to the rest of the animal, termed 'limb girdles'. In the case of the forelimbs this is the bones of the shoulder (or pectoral girdle), while the pelvis (or pelvic girdle) serves this function in the hindlimbs.

The shoulder begins with the scapula: a long, thin bone that lies across numerous dorsal ribs (it is partly anchored to them, thus ultimately holding the arm in place) and runs down to the chest. Here a smaller bone, the coracoid, links to the scapula, and between these bones lies the joint for the humerus, with the scapula and coracoid also providing attachment sites for the arm muscles. Linking the two shoulders in the middle of the body is a bone called the furcula, which is composed of fused collarbones (better known in birds as the wishbone).

The furcula has had a long and troubled history in dinosaur research. In the early twentieth century a seminal book on the origins of birds thoroughly assessed the information available at the time, and concluded that the theropods had more in common with birds than with any other animal group, but then summarily ruled them out as bird ancestors on the grounds that they lacked a furcula. The prevailing idea at the time was that once a structure was lost it could not be regained, and other reptiles had collarbones while dinosaurs did not. Thus if the dinosaurs lost theirs, they could not have given rise to birds, which did have a furcula. This idea was generally incorrect (lost features can certainly return), but it was also due to a data problem. With more time came better dinosaur skeletons, and we now know that numerous dinosaurs, including tyrannosaurs, had a furcula.[11] More than this, however, the furcula sits between the coracoids, and in tyrannosaurs this means that the arms were positioned close to the middle of the chest, not on each side of the body as is often seen in old reconstructions.

Moving down the arm, there is first the humerus, then the paired radius and ulna that make up most of the arm (Fig. 11). These bones are largely unremarkable in comparison to those of other theropods, although they shorten markedly over time and are proportionally short in the later tyrannosaurs. They are not reduced to the point where they lack a function, however, they retain major muscle groups and articulations, and would still have been quite strong despite their short length. Theories on the function of the forelimbs in derived tyrannosaurs have been extremely controversial, and some of the hypotheses are discussed in more detail later.

The hands are also rather unspecialised for theropods, with a series of small wrist bones, then the long bones of the hand (the metacarpals), then the finger bones (the phalanges). Most animals, including tyrannosaurs (but not humans), have different numbers of bones in each finger. Humans are also somewhat unusual in having five fingers on each hand – although

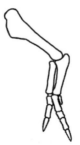

*Fig. 11 Early tyrannosaurs had relatively long forelimbs but the arms of later species (here the left arm of* Tyrannosaurus*) were proportionally short and with only two fingers with a relatively weak grip.*

this was the original number settled on by the early ancestors of tetrapods, in a great many groups the number of fingers was reduced later on. The earliest theropods had four fingers, most had three, and famously *Tyrannosaurus* and the other tyrannosaurines, alioramins and albertosaurines had just two. Several specimens have an extra metacarpal (not an extra finger as is sometimes reported), so there may be three long bones in the hand but only two fingers. The extra split of bones seems to come and go, suggesting that this feature was not under any strong evolutionary selection and was essentially vestigial. The fingers are a little more unusual than those of most theropods, which have a rather robust and slightly offset first finger; while it was hardly a thumb like that of a human, it was at least a little more like this than the other digits. This arrangement applies to animals like *Guanlong*, but in derived tyrannosaurs the two fingers are similar in overall shape and proportion, and the first finger is not offset.

The identity of the fingers in theropods has perhaps been an area of more controversy than any other. In birds the fingers that form in the embryo are II, III and IV (in other words, the middle three fingers of the ancestral five-fingered tetrapod hand, with I being the thumb) – though these later reduce and coalesce, and most birds have only two fingers as adults, and even these are often nubs. This is largely what you would expect: when fingers are reduced or lost in animals the first to go is inevitably the fifth finger, followed by the first. However, in theropods, although the fifth finger is lost first (as can be seen by the presence of four fingers, then a stumpy metacarpal in very early forms), the next one to go seems to be the fourth. This implies that theropods leading up to birds had fingers I, II and III, which is different from the avian II, III and IV.

This conundrum is occasionally used to argue that birds are not descended from theropods, but there are some possible explanations for it. First of all, there may have been what is termed a 'frame-shift', where essentially the developmental programme of the hand moves over one finger, so the second digit develops like a first, the third like a second and so on. Such developmental phenomena do occur in nature, and all kinds of developmental anomalies occur in digits. They may be seen, for instance, in people with extra fingers, or with 'reflected' hands with a thumb on each side of the hand. Such a development may well have occurred at some point in the theropod or early bird tree, though it would be impossible to detect unless we found some perfectly preserved dinosaur embryos of just the right age (which is not impossible). Another explanation for the development could be that the theropods did indeed have digits II–IV, and lost digit I, while digit II changed into something more like digit I. An odd early ceratosaur called *Limusaurus* from the Middle Jurassic of China had a rather thumb-like digit II, and a much-reduced digit I.[12] This would suggest that an evolutionary change may have occurred relatively early on in theropod evolution, and that digit I was indeed lost, so that theropod hands would match the pattern seen in birds, of II–IV digits.

Regardless of which of these (assuming it is either) is responsible for the discrepancy between most theropods and birds, this has obviously caused problems when describing the fingers of animals like tyrannosaurs: did they have a I–III or II–IV finger pattern? Convention remains with the former, while the two-fingered species would have lost digit III as well, and have just digits I and II. As already noted, the fingers have varying numbers of bones in them. In tyrannosaurs digit I had two bones, II had three and III had four. The first digit was rather shorter than the others, but the second and third digits were of a similar lengths – there is an extra bone in the third finger, but it is very short and adds little to the length. These values include the last bone of the finger, which is specialised and forms the underlying shape of the claw, and is called the ungual.

The unguals are hook shaped, and form and support the keratinous claw (discussed in more detail later). The bone itself gives an idea of the shape of the complete claw, but it may not have had as close a relationship with the external claw as might be imagined, and there can be a fair bit of difference between the two in terms of both overall size and exact shape. Some claws were probably considerably larger and more curved than you might expect based on the underlying ungual alone. Grooves run down the sides of the ungual and to the tip of the claw, though their function is not clear. It has been suggested that they may have been anchor points for the keratin, or that blood vessels may have sat in these grooves, supplying blood to the keratinous claw sheath to help it grow.

When it came to the arm joints, the shoulder allowed a fairly free rotation of the arm, but the other joints were more restricted. As in humans (and indeed most animals), the elbow was a fairly simple hinge and allowed the lower arm to fold up towards the humerus. However, rotation at the elbow was limited in dinosaurs generally, and particularly in theropods the hands would face each other as if they were about to applaud. Numerous reconstructions get this wrong, showing the hands with the palms facing down and with strongly bent wrists (often called 'bunny hands'). Many later theropods did have wrists with a great degree of flexion, but in a different plane, as if trying to get their outer fingers to point back along the ulna to the elbow. This was a key development for the flight stroke in birds, although the tyrannosaurs had rather more limited movements here.

The fingers could each curl up as in humans, and each joint allowed a fair degree of flexion. However, the fingers of tyrannosaurs had fewer elements than those of humans, and with each finger having differing numbers of joints, tyrannosaurs would not have been able to make something quite like a fist with the hand. Early tyrannosaurs possessed good flexibility in the joints and a strong grip.[13] This is evidenced by the deep pits on the sides of the phalanges that anchor the ligaments attaching to the next bone in the series. The deeper the pits, the more strongly the ligaments were attached, so the stronger the grip was. In time, the pits became more shallow, showing that later species had a weaker grip. The base of the ungual is often expanded to give an attachment for muscles and especially ligaments, providing strength in this area and allowing the claws to puncture whatever was in the hand (potentially prey).

## Longshanks

The second girdle is that of the pelvis, consisting of the three major elements, the ilium, pubis and ischium, and the first of these is fully fused to the sacrum in adults to provide a structurally rigid framework for the legs. Tyrannosaurs are rather odd in having 'pinched' ilia: when seen from above they appear to almost meet in the midline about the neural spines of the sacral vertebrae, giving them a characteristic appearance. Quite why this is is not clear; this feature turned up even in early forms like *Guanlong*, at a time when the animals were still small and had less specialised legs than those of the later giants, yet it was retained and even exaggerated in the later tyrannosaurids. The ilium does provide a site for muscle attachment for various muscle groups that move the femur, especially to move it forwards during locomotion, though as previously noted the key retractors are those connected to the base of the tail.

The rod-like pubes and ischia on each side meet in the midline under the body and fuse, with the pubes to the front and the ischia behind. The ilium supports several muscle groups used to retract the femur, while the pubis helps support the body and also anchors some other muscles groups in the legs. In the later tyrannosaurs the end of the pubis became extremely enlarged, while the ilium reduced a little − though again, the reasons for the shifts of these features are unknown. Where they meet, the three pelvic bones all contribute to make up the edge of the acetabulum, which is the big hole in the pelvis that takes the head of the femur. Actually, this hole is rather larger than the head of the femur, but it would have contained a lot of cartilage to prevent the bones from grinding together as the animal moved. The top of the acetabulum is a shelf-like extension of the bottom of the ilium that helps to hold the femoral head in place.

The femur itself is somewhat typical for theropods, being an inverted L shape. This gives it a long shaft with a hemispherical head that is close to perpendicular to the shaft, and effectively plugs into the acetabulum. The head provides articulation and allows the femur to move, while the straight shaft helps to take the weight of the animal. This arrangement in the bone is exactly what you would expect in an animal with an upright stance (that is, with the legs directly under the body), and is also seen in birds and mammals; it is different in sprawling animals like lizards and newts, in which the weight does not bear down on the long axis of the femur. It is one of the key features that shows that dinosaurs were generally upright animals, not lizard-like sprawlers, and it is clearly visible in the tyrannosaurs.

Moving down the limb there is the large tibia and the rather smaller fibula. In birds the fibula is almost non existent (as in later pterosaurs), and some of the ankle bones are fused to the tibia to make a complex lower leg rather more simple, but tyrannosaurs, like other non-avian theropods, retained these bones as separate elements. On the front face of the bottom of the tibia there is a large recess for the astragalus. This is the largest bone in the theropod ankle. It has a huge triangular piece that sits in front of the tibia, giving theropods' lower legs a characteristic appearance. As in the wrist, there is a series of bones in the ankle. The base of the astragalus is rather like a roller joint, allowing the foot to swing forwards and back, but it is otherwise quite rigid – unlike humans, tyrannosaurs would not have been at much risk of a twisted ankle.

Below are the five metatarsals. The first, equivalent to our big toe, is very small and is positioned halfway down the second metatarsal; it is a tiny triangular nub that supports a small toe of just one phalanx and one ungual. This is essentially equivalent to the dewclaw that is possessed by many animals, and is a greatly reduced first digit. The fifth metatarsal is even more reduced – there is no toe at all, just a short, slim and slightly curved bone on the outer part of the foot. With the first toe being much reduced and the fifth all but non-existent, the foot therefore has three functional toes for walking and so therapods leave three-toed footprints.

The main metatarsals are long and straight pillars, with the outer bones (II and IV) diverging slightly at the end so that the toes spread out from one another. In later tyrannosaurs, however, the foot is rather modified – the central of the three metatarsals has a very slim proximal part, and is 'pinched' between the normal bones of the second and fourth metatarsals.

This condition stabilises the foot, as the middle metatarsal is restricted in its movement between the metatarsals on either side when walking, or especially running. A lot of energy is lost to this kind of movement when moving quickly, so this adaptation helps make an animal a little faster, but more importantly makes it more efficient.[14] It would enable an animal to use less effort to travel long distances, something that fits with what we have already seen with the longer femur in tyrannosaurs, and long metatarsals generally; tyrannosaurs had a long stride and an efficient walk.

The metatarsals, however, would not normally contact the ground. Humans are rather odd in that we walk on the whole foot with our toes and metatarsals forming the contact with the ground. However, tyrannosaurs were like most animals that walk only on the toes (or in extreme cases, like in horses, on the tips of the toes), using the metatarsals to effectively increase the length of the leg.

Like the fingers in the hand, each toe has a different number of phalanges and ends with an ungual. Including the unguals, there are three elements to the second toe, four in the third and five in the fourth. Again mirroring the hand, although the toes have different numbers of bones in them, they are about equal in length – though the third toe is a little longer than the others. The unguals are largely conical in shape and not anything like as curved as those on the hand; in short, they are there primarily to get to grips with the ground rather than anything else.

## Walk this way

Determining how the tyrannosaurs moved requires a complex interaction of different fields. We can reconstruct the skeleton and the major muscle groups of a tyrannosaur with some accuracy, and an understanding of the overall build of these animals (including issues like the position and size of the lungs and air sacs) shows where the centre of mass lay. In order to balance, a moving tyrannosaur needed to keep its centre of mass almost over the feet or it would have toppled one way or the other. We can then begin to work out the pattern of movement with computer models of the animals, and combine this with data from footprints.

Individual dinosaur footprints and sets forming tracks are quite common, but the problem is working out which species a given track might relate to. Tracks vary depending on the animal leaving them, and also on how it was moving, what the surface was like when it left them, and any changes like erosion that happened afterwards. As a result, there are plenty of tracks that cannot with certainty be distinguished between those of a hadrosaur and theropod, and pinning them down to a clade let alone a genus is all but impossible. Even so, there are a couple of tracks that we know were definitely made by tyrannosaurs,[15] as they are obviously huge tracks of three-toed theropods, from a time and location where tyrannosaurids were the only big theropods around.

Of course, even knowing that you have a track of a tyrannosaur doesn't get you that far in identifying it given the issues noted above. Still, there are at least a few relatively consistent relationships between the size and shape of a dinosaur's foot, spaces between tracks and how they are moving. The length of the foot gives a good indication of the size of an animal, and the spaces between the tracks show how long the stride was and are therefore linked to speed. Add that to our models of movement and how the muscles work, and we can build up a picture of acceleration, turning and top speeds for tyrannosaurs, although exact values are always going to be rather uncertain.[16]

As to the actual pattern of footprints, well, obviously tyrannosaurs were bipeds and they would have moved one leg at a time. This might seem obvious, but although quadrupeds have a remarkably diverse pattern of footfalls at different speeds and times (compare, say, a giraffe walking, a horse at a canter and a cat walking), even the bipeds can move in different ways. Small birds, various rodents like gerbils, and kangaroos and wallabies hop using their back legs in unison, while some animals do a sort of 'skip' with one short step followed by a long one leading to an odd asymmetrical run (vultures do this quite a bit), so tyrannosaurs actually walked pretty normally.

The femur is rather long in tyrannosaurs compared with the femur in other theropods, but the lower leg and metatarsals are proportionally even longer. An animal that runs quickly often has a short femur, since in order to take a full step the femur must be brought though an arc. If the femur is short you can get through more steps more quickly. The loss in stride length resulting from a shortening of the femur is made up for by a long tibia and/or long metatarsals. Look at the leg of a horse or greyhound, and for all the length of the leg and the speed these animals achieve, the femur is really rather short and makes up little of the length of the leg. In a tyrannosaur the femur is long, so is perhaps not capable of a very high number of strides, but this would be offset by the longer than usual lower parts to the leg, which indicate that tyrannosaurs would still have been relatively rapid walkers or runners.

At all speeds tyrannosaurs appear to have travelled one leg at a time, and though larger forms would have been too large to run, they could still move pretty rapidly: a long stride length can make up for a lot. In a normal standing position, the feet would be side by side, and the femora and tibiae would be almost vertical and form a column to the ground, but when walking these bones would be closer to the midline of the animal. With each step the leg would swing a little under the body and plant the foot under the hips, rather than to the side, thus taking the weight of the animal more effectively. Watch how a chicken walks, or better yet an ostrich, and you will see that its toes retract with each step so that they can slip past the standing foot, then splay out as they are placed below the body. Looking at bird and theropod tracks (and even the tracks of humans), it's fairly clear that the legs don't just move from the front to the back, but swing in as well, and this characteristic becomes more pronounced at higher speeds.

The information we have gives us a good idea of how tyrannosaurs could use their arms and legs. Putting this into context with their other anatomical features, and in particular with the environments they lived in and the other animals they co-existed with, allows us to piece together how they lived. The anatomy of an animal provides a great deal of information about its function, typical actions, abilities and limitations, and thus about how it was likely to have lived (and indeed died) – even if it has been extinct for tens of millions of years. This, then, is the next step in reconstructing a real tyrannosaur – but first we must complete the animal with its skin.

# CHAPTER NINE

# Outside

Having dealt with the insides of the tyrannosaurs, we now come to the outside. All organisms need some kind of barrier between themselves and their environment, to keep their organs in and protected, and to keep out any pathogens (such as bacteria and viruses). In the case of vertebrates, the barrier is the skin. A key development of tetrapod evolution was the formation of skin containing a considerable amount of keratin, which made the skin of early ancestors of reptiles impermeable and enabled them to live out of water. Unlike modern amphibians, reptiles don't dry out when away from water, and this allows them to live in dry environments. Although scales were lost in the mammal branch of the amniote tree, and reduced in some reptiles (and of course birds), most reptiles, including the dinosaurs, were scaly.

This was obvious to early palaeontologists, who identified dinosaurs as reptilian. Some preserved footprints and impressions of dinosaur skins also showed that they had scales, as do modern lizards, tortoises and crocodiles, and for the next 200 years illustrations of dinosaurs showed them to be scaly. Oddly, even when it became apparent that some theropods were very close related to birds, and that *Archaeopteryx* and the earliest birds were fully feathered – and with rather specialised feathers at that – researchers and illustrators alike seemed reluctant to ascribe feathers to anything else. Even though it was clearly unlikely that feathers would have sprung up as a full body covering on the first birds from fully scaly dinosaurs, this was more or less what was surmised.

However, in recent years spectacular feathered dinosaurs have come out of China, Germany, Brazil and Canada, and from these we now have an exceptional idea of the origins and evolution of feathers in the Dinosauria, before the existence of birds. Feathers, or at least feather-like filaments, may in fact predate the dinosaurs and could have originated in the archosaurs, before being lost in the sauropodomorphs and most ornithischians (several species of the latter are known to have had them, as are many pterosaurs), but true feathers were certainly present in the coelurosaurs, including the tyrannosaurs.

We have direct evidence of feathers on two tyrannosaurs: the small *Dilong* and the much larger *Yutyrannus*.[17] Both show evidence of possessing simple, down-like filaments that are rather like those of baby birds. Feathers are evolutionarily derived from scales, albeit very modified ones, and feather themselves have evolved into a variety of very different forms. Birds (and indeed various non-avian dinosaurs) have a wide variety of feather types, from the down we see in modern chicks, and long filaments in kiwis and cassowaries, to the more 'traditional' feathers like those on the bodies and wings of flying birds. Feathers have been greatly modified in various lineages, and there are some real oddities in various groups of theropods. The curious scansoriopterygids (often thought to have been early flightless birds, but perhaps likely to have been very odd early oviraptorosaurs) had feathers that were solid sheets of keratin; one therizinosaur seems to have had quill-like feathers that might have functioned in defence as do the quills of porcupines and hedgehogs, while penguins have heavily modified feathers designed for an aquatic lifestyle. Only later in the evolution of dinosaurs do more modern-type feathers with a central shaft and branches from the shaft appear. In tyrannosaurs and compsognathids the feathers were limited to simple filaments, some of which could be quite long – in places, those of *Yutyrannus* were 10 centimetres or more in length.

The preservation of feathers in the only known specimen of *Dilong* is patchy, but two specimens of *Yutyrannus* (preserved together as a pair) show that in this animal at least, feathers covered most of the body. They are seen only in an outline around the edges of the specimens (which are rather flattened and not 3D like most tyrannosaur fossils), but it seems unlikely that the feathers would have run over the back, belly, arms, legs and even the head, and not covered the flanks, too. In overall appearance the feathers would perhaps have looked more like fur or long hair than anything else. Feathered dinosaurs like this are often depicted as being rather shaggy, and while this is not impossible, most animals with fur or simple filaments have a neat and trim appearance and are smooth to look at: the average mammal has a coat more like that of a Labrador than of a Yorkie.

Feathers are rarely preserved as the conditions for their preservation have to be near perfect, and we have to be lucky enough to find fossils of feathered specimens. Thus even though only two tyrannosaurs are directly known to have had feathers, it is reasonable to infer that all of them did. Feathers appear around this point in the theropod tree, and everything from the compsognathids through to modern birds has at least some representatives in the fossil record with feathers; we also know that none of these animals lost them over time: once feathers turn up, they seem to stay. Thus, perhaps sadly, all those iconic images of a scaly *Tyrannosaurus* battling a *Triceratops* are almost certainly incorrect. It was once suggested that juveniles of the larger tyrannosaurs might be fluffy, and that the adults were bald and scaly, the idea being that larger animals would overheat if insulated (this is part of the reason why large animals like elephants and rhinos have shed their fur). However, the full feather coverage on *Yutyrannus* (an animal that weighed well over a tonne) would seem to counter this line, so we might expect even adult *Tarbosaurus* and *Gorgosaurus* to have had a good covering of feathers.

It is notable, however, that the feather coverage need not have been uniform. Although most birds are covered from the head down to most of the feet in feathers, this is not the case in all of them. Ostriches, for example, have little feathering on their necks, which are largely bare (hence appearing pink), and their thighs are also largely devoid of feathers. This may be to help prevent overheating: when moving at top speed the muscles in the legs have to work extremely hard and generate a lot of heat. A thick cover of feathers would keep in the heat and perhaps cause problems, so maybe tyrannosaurs had strategic bare patches that served to regulate their temperature more effectively; this may have been the case especially in large forms living in warm environments.

We tend to think of feathers as being primarily 'for' flight, but the fact that they turned up millions of years before the appearance of the first avians suggests that they must originally have had some other purpose, or multiple different functions. Indeed, feathers have been adapted to fulfill a huge number of functions in modern birds. They can be waterproof and help buoyancy in water, insulate and keep in heat, and act as display or warning signals; they can be used in the brooding of eggs and keeping chicks warm, as well as for defence and other purposes. Display and insulation appear to have been the two main drivers in early feather evolution in theropods: they may have provided benefits to feathered versus non-feathered animals, and thus driven selection for their propagation. Certainly, feathers have the advantage over scales in that they can easily be shed, which makes it easy for an animal to change coats from, say, summer to winter, or to alter its colours for the breeding season.

Feathers can also take on some fairly specific roles with a little modification. Several bird types, like nightjars, have elongate feathers around the mouth, termed 'rictal bristles', which appear to function as something of a sensory device, a little like the whiskers in a cat. Others have something analogous to the mammalian eyelash: bristle-like feathers that help to protect the eye from dust and the like (these are really obvious in the larger hornbills). The former might be a bit of a stretch in a multi-tonne tyrannosaur, but it is easy to imagine that some of the plains- and desert-dwelling tyrannosaurs may well have evolved a set of eyelash feathers to shield their large eyes from sand and grit.

Scales themselves are of course fundamental in non-avian reptiles (though there are some captive-bred snakes and lizards that lack scales, just as there are some scale-less fish and feather-less chicken breeds), but they vary in size, shape and distribution. Like feathers, and for that matter mammalian hair, and fingernails, claws and even beaks, scales are composed of keratin, which is pretty tough stuff (this is partly why it tends to survive the early stages of decay to be fossilised compared to things like internal organs), a characteristic that is desirable in something that effectively acts as a barrier to the outside world. There is a fair bit of keratin even in unadorned skin such as our own.

It is not clear what causes the variation in scale patterns in most reptiles, though in some cases the reason for it is relatively apparent. In crocodiles the scales are on top of pieces of bony armour in the skin, while in tortoises they cover the shell. In some places (like at the back of a knee) scales require a lot of flexibility and may be very small or sparse; in others they may be large or small or a mixture of both, closely packed together or far apart, arranged in lines or radiating patterns, or laid down apparently at random. As a result, without good preservation it is hard to predict the patterns that may have been present in various dinosaurs, and for the tyrannosaurs we essentially have nothing. Although in some feathered dinosaurs the feathers covered the fingers and toes, the feet especially seem to have been rather sparsely feathered, so these are areas that may have retained scales even in an otherwise very feathery tyrannosaur. The snout or part of the head may also have lacked feathers, but quite what the distribution of feathers would have been beyond that in the average tyrannosaur is unclear.

There is a tendency to think that scales and feathers are mutually exclusive – that if there were feathers, there were no scales – but there's no particular reason why this should have been the case. It's clearly not impossible for there to have been different types of structure on the same animal. However, things are more complex than this. For a start, scales are rarely preserved in fossils, and even when feathers have been preserved, skin and scales may not have been, so their absence doesn't necessarily mean very much. Even fossil birds with near-perfect feathers appear to have nothing on their feet, when in fact we'd expect them to have scales (either dinosaur or avian types), but they have simply not been preserved.

More importantly, scales don't necessarily exclude feathers. Some birds have small filamentous feathers on their feet that poke out from between the scales (and these may go right down to the toes), so just because there are scales, feathers should not be ruled out. As it happens, though, the scales on the feet of birds are not really the same as those on other reptiles, because birds went through a phase during which they lost all their original scales, and the scales seen in modern birds are actually a secondary redevelopment of scales from feathers. Evolution can take a roundabout route sometimes, though it seems scales on the feet are likely to offer some protection to them. Thus, although we can be confident that tyrannosaurs had a liberal coating of feathers, it is not impossible that there was also considerable coverage of scales in places (even in areas other than the feet), and feathers may well have intermingled with scales.

## Beyond scales

Feathers and scales are not the only possible outer covering in an animal. In the tyrannosaurs areas of skin may have lacked both – the bald heads of vultures and the necks of some tortoises and terrapins, for example, lack any other kind of cover – and they may have had something that was much more like the skin of mammals in general appearance. It may have had some thickened keratin on it, too, to make a feature larger or provide it with better protection. The face may have lacked scales or feathers, but could have had any combination of cockscomb crest, wattles, dewlaps and other features that would not show up in the fossil record except under the most exceptional preservation conditions. This is not to say that tyrannosaurs *were* adorned in this way, but speculative reconstructions giving them a large mass of rooster- or turkey-like soft tissues on the head and neck, or even more extreme features such as the inflating sacs of tropicbirds, or coloured throats like those of tragopans, are no less possible than a lightly scaled or feathered face.

Given their carnivorous lifestyles, you could be forgiven for thinking that tyrannosaurs may have had bald heads and necks (like vultures), which would have been easier to keep clean than feathered features when dealing with bloody carcasses. They might well have lacked feathers on the face, but I doubt that it would have been for this reason. Vultures stick their heads deep into the body cavities of dead animals, but tyrannosaurs would probably have either taken small prey, or broken up and consumed parts of large prey, rather than rummaging around inside, so this would not have been as much of an issue in them as it is in vultures. In addition, many other carnivores that feed messily, like lions and hyenas, don't have bald heads either, so this is clearly not a critical adaptation for this kind of lifestyle.

Rather than having bald heads with bare skin, tyrannosaurs may have had a gnarled mass of keratin. As mentioned in the skull section earlier, this may have been the case especially in the areas around the snout, where the little projections and dimples on the nasals would have provided an ideal support for an extra layer of keratin, and we would expect such extensions on the hornlets and bosses over the eyes. Some reptiles have something like this over horns and parts of the head (a number of iguanas and chameleons, for example, have such features), so it's feasible that dinosaurs also had them.

One issue that has been hotly debated recently has been the question of whether or not many theropods had lips. The mandible of a theropod is generally more narrow than the skull, so when the jaws are closed the teeth of the lower jaw go behind those of the upper jaw, giving the appearance of an overbite. As a consequence, theropods are generally illustrated with the teeth of the premaxilla and maxilla 'on show', forming a ring around the snout and giving the impression that they were clearly visible in the living animals. However, theropods may have had lips that covered the teeth when the mouth was shut, thus making them look more like lizards than crocodiles. In the absence of any obvious bony feature that could correlate with a soft tissue structure, we generally look to the nearest relatives for information, but doing so is problematic in this case.

Birds obviously have beaks rather than teeth, so are probably not a good reference point. Crocodilians do not have lips at all, exposing all their teeth, and indeed this aspect of their appearance may well be what drove the conventional picture of toothed theropods. However, modern crocs and their relatives live in water, and when striking at prey and feeding they need to be able to easily drain out all the water from their mouths; a fleshy lip would trap water, slowing down the bite and probably costing them their dinner. Thus they may also be a poor choice for working out what the theropods did. If we turn to other living reptiles that lack beaks — lizards, snakes and the tuatara — it can be seen that they universally have lips. Even in large predators with very elongate teeth and something of a theropod-like overbite, such as the Komodo dragon, not a single part of any tooth is visible when the mouth is shut. Again, it is hard to be exact without a specific example from the fossil record, but it seems that experts are increasingly leaning towards lipped theropods, including tyrannosaurs. That famous 'ring-o'-teeth' look could be on its way out.

One final note on the general appearance of the dinosaurs relates to how the skin fitted over the body. In various incarnations in dinosaur illustrations, there has been a tendency to shrink-wrap the animal: effectively to put on almost the minimum of muscle, then coat it in skin like a layer of latex. In areas like the head or feet, where there would have been little muscle on top of the bone, dinosaurs were generally shown as having a near-skeletal appearance. This was partly because some artists tried to show the skeletal remains while creating a reconstruction of the animal, an odd but understandable and sometimes useful technique, but one that became something of a meme, with some animals coming across as little more than skeletons with skin.

We do now have a better understanding of the sizes and distributions of muscles, which has helped. We also have more of an appreciation of biology, and know that there would have been deposits of fat in places, and that in the case of the skull, for example, there would have been at least some generic connective tissue between the skin and bone. You can't immediately see the details of the skull bones and fenestra in most lizards and crocs through the skin, so why would we expect this in dinosaurs? Tyrannosaurs would also probably have had various fat deposits on the body, further separating their outer shape from the skeleton, and some might have fed heavily for a few months and starved in other months, much like brown bears do, and as a result alternated between the two extremes of plump and skeletal in single animals (another reason why mass estimates can be so variable).

The final external part of tyrannosaurs to deal with is the claws. The bony unguals described earlier were covered with a thick layer of keratin to form sheaths. Incidentally, in humans and other primates, unusually this sheath has been reduced until it is buried in the skin, forming a fingernail rather than a claw that covers the length of the last bone of the finger. The shape of a claw does have some links to its function: claws that are particularly curved are good for gripping, for example, but there are problems in interpreting them. First of all, although there is a general relationship between the bone and sheath, the shape of the sheath is oddly not always a good match for the shape of the bone that supports it.[18] The bone is usually all we have to go on, but the shape of the sheath is what helps elucidate its function, so we have a mismatch that makes it hard to decipher the function of any given claw. Second, even when sheaths are preserved, they are generally in 2D, so we don't know if the underside edge was rounded or sharp. In side view a knife blade can have a similar profile to something like a nail (one is good for puncturing, the other for cutting), but without knowing if there is an edge there it is hard to work out what function a claw might be better suited to from the ungual alone.

The pedal (foot) claws of tyrannosaurs are not strongly curved. Since they would have been used primarily for walking, and the claws of terrestrial animals generally have near-flat undersides, this matches what we would expect. The claws would be growing continually, since they would wear down with use, but the tips may still have been relatively sharp. Though the claws were not built like the talons of birds of prey, a hit from a tyrannosaur foot might still have been pretty nasty. The manual claws are more curved and thinner, and thus better set for gripping. This makes sense in early tyrannosaurs, which had larger hands and longer arms than later ones, and their claws may have had a role in prey capture. In later forms that had small hands, reduced fingers and short arms, this would not have been the case; the reduced ligamentous pits on the fingers correlate with this, implying a lack of a strong grip and reduced use of the claws.

In short, not only were tyrannosaurs not scaly, but in fact they would probably have possessed a mosaic of different covering types – of scales, feathers, unadorned skin and keratin – on various parts of the body. Unsurprisingly, the covering would have varied depending on the different demands on different parts of the body, and would have reflected such demands. The feet would have had thick scales for protection, and the snout probably had an extra layer of keratin for the same reason. There may have been no scales behind the knee or between the finger joints, and parts of the body may have had few filamentous feathers to help prevent overheating, while others might have borne longer filaments for display. We currently know little about this, though we at least have a good idea of what might be expected, and there are good reasons for these inferences. Increasing amounts of skin and feathers are turning up for various dinosaurs, including tyrannosaurs, and this is likely to be a productive area in the near future.

# CHAPTER TEN

# Physiology

W e can assume that most of the basic functions of the tyrannosaurs were the same as those of other vertebrates – they had to breathe, metabolise food using oxygen to generate energy, their muscles contracted to produce movement, they excreted waste products and so on. The basic physiological properties of the tyrannosaurs are thus not covered here, and the focus is exclusively on one critical aspect of physiology – their levels of activity and ability to regulate their temperature.

The subject goes back to the earliest research on dinosaurs, when there were disagreements between Richard Owen, who regarded dinosaurs as giant lizards, and Gideon Mantell and Joseph Leidy, who thought of them as more active animals. This fundamental division has continued in scientific circles, and it encompasses a vast range of controversial ideas and analyses that have been bandied back and forth between palaeontologists and biologists for decades. Did tyrannosaurs have a fundamentally low metabolism, reliant on the heat of the sun or the general temperature of the environment, in order to be active; or were they capable of generating their own heat and thus able to be active at any time of the day, and to engage in sustained high-energy activities?

## Blowing hot and cold

In short, the question often asked is simply: were the tyrannosaurs warm- or cold-blooded animals? An attempt at an answer requires an understanding of metabolic rates and how we might be able to detect them in tyrannosaurs, and would one answer apply to the whole clade or even to individual animals?

The division into hot- and cold-blooded animals persisted in scientific circles for a long time, and this is the main reason why it has hung on in the public consciousness. It is a simple and convenient shorthand (probably the other reason why it has survived) for active versus inactive, but these days much more nuanced descriptions are required.

Traditionally, the mammals and birds are considered to be hot-blooded animals, and in fact because of this shared feature the two groups were once thought to share a common ancestry. It is, after all, rather special and is not seen in any other animals, so for all the variety of insects, molluscs, crustaceans, fish, amphibians and numerous other groups, only these two are apparently warm-blooded (though recently it has been discovered that at least one fish, the wonderfully named comical opah, is warm-blooded, at least compared with other fish). Modern birds and mammals do indeed typically have hot blood: they are warm to the touch and, more importantly, stay this way even in cold conditions and over extended periods of time. They produce their own heat and keep themselves warm, even when it's cold around them.

This is quite a trick: it allows them to live in places where other animals cannot normally survive (how many lizards live in the Arctic Circle?), and to be active at times of the day or year (during cold nights and winter) that other animals cannot manage. The cost, however, is considerable, because to produce the heat, they need a lot more food than do comparably sized animals that are not 'warm-blooded': in fact somewhere between six and seven times as much. That's why zoos feed their lions every couple of days, while a large alligator or python is fed only every week or two.

Cold-blooded animals need just a fraction of the calories of warm-blooded ones, but can also survive for long periods without eating or doing anything very much; records of lizards surviving anything from six months to a year without food are quite common. On the other hand, cold-blooded animals are limited to certain locations, habitats and conditions, and cannot easily 'get going' if the temperature is not high enough. They typically obtain heat from the sun (that's why they are limited in winter and at high latitudes), and spend time basking to soak up the sun's heat, which also makes them vulnerable to predation.

But why go to all this trouble to stay warm? Why do animals need to be warm in the first place? It all comes down to the action of physiological processes, and especially enzymes. These proteins carry out huge numbers of chemical reactions to keep organisms operating, and they function most efficiently at temperatures of 30 or so degrees centigrade. Animals maintaining such a temperature can therefore be more active more consistently, as their enzymes are at the optimal temperature. This system is particularly efficient in large organisms, since with increases in size, the volume gets bigger much faster than the surface area does. A cube with sides 2 centimetres long will have a volume of 8 cubic centimetres and a surface area of 24 square centimetres, but if you increase that cube to 20 centimetres on a side the volume will be 8,000 cubic centimetres, with the area being just 2,400 square centimetres. Therefore any heat inside a large organism will take longer to leave and it will be warmer for longer, but smaller animals will struggle to keep the heat in. This may be part of the reason why few lineages have produced warm-blooded animals: in small animals the heat would bleed off too quickly, and there are proportionally few species that weigh more than a few grams.

There are various ways of getting warmth and keeping the temperature at a fairly high level. It's not simply a question of 'being warm' as a result of being a mammal or bird, and producing heat from physiological processes. Physiologists separate out the difference between having a high temperature or not, and also how that temperature is achieved and maintained. So rather than simply using the terms 'warm' and 'cold blood', they use the words endotherm and ectotherm, and heterotherm and homeotherm. 'Therm' relates to temperature, with the prefixes 'endo' and 'ecto' meaning inside and outside respectively, and 'hetero' and 'homeo', meaning different and similar. Therefore endotherms are animals that get their heat from inside, and ectotherms get it from the environment, while homeotherms maintain a constant temperature and heterotherms do not. It might seem that these terms are redundant – surely endotherms are also homeotherms? – but this is not the case. Animals have evolved a variety of different metabolic patterns, and there are interesting exceptions and alternatives that show just how variable things can be. Animals can change their physiological patterns over single days, over years or over an entire lifetime, and they vary according to both size and activity.

For a start, a number of mammals are not that good at maintaining their own temperature, and cool down and even die if the environment is not warm enough. The platypus and naked mole-rats both need environmental heat, so are best considered ectothermic homeotherms: they have a stable body temperature, but it's at least in part supported by the outside world. The temperature in other animals drops dramatically when they rest, whether it's for the long haul as in hibernating bears, or overnight as in birds like hummingbirds, so they are considered heterothermic endotherms: they generate their own heat, but their body temperature is variable.

Moths are an especially interesting case. It is often suggested that only endotherms have insulation as they need to keep in as much heat as possible given the energy they have expended to generate it, but moths are largely furry, even though as insects they are very much ectothermic animals. However, moths buzz their wings furiously before flying and this activity generates a lot of heat in their muscles, warming the animals through, while the fuzz helps to keep in the heat and more flight continues to generate more heat. Moths can operate, at least for a while, in low temperatures (at night) by keeping active and insulated, making them ectothermic heterotherms.

At the other extreme, both large sharks and tuna can exploit the surface area-to-volume ratios in their favour. Both are ectothermic homeotherms: they maintain a stable body temperature, but like moths, generate heat from their muscle activity (this counts as being ectothermic, even if it is internally generated, because it's rather different from having a higher basal metabolism, and if they stopped moving they'd cool down). This allows them to remain constantly active by keeping their temperature up, and they are always in a position to engage in major activities: quite an advantage in cooler waters where other fish would be more sluggish.

Obviously, many ectotherms get their heat from the sun by basking, and if the environmental temperature is high enough they stay warm. This is why there are plenty of nocturnal lizards and snakes in the tropics, where the nights are balmy, but in cooler climates they are generally limited to the day when the sun is out. Crocodilians can further exploit the water to the same effect: water cools down rather more slowly than air, so crocs can take a dip and pick up some heat in the night, or alternatively cool down in the day when the water is cooler than the air. Big crocs in particular also get the benefit of having a large volume and small surface area, and can stay warmer for longer more effectively as a result, so they could be lumped in with the sharks as ectothermic homeotherms.

Finally, it must be remembered that large endotherms can actually struggle for the same reason: they can overheat. Asian elephants that have the options of forest shade and water are rather hairier and have smaller ears than their savannah counterparts in Africa, which are both larger and exposed to the tropical sun, so are nearly bald and use their ears to cool down. Rhinos and hippos are also largely hairless; size and heat combined can be a problem for them. Collectively, therefore, there is a wide range of evolutionary and behavioural strategies animals can adopt, with a real mixture of endothermy, ectothermy, homeothermy and hetrothermy to consider, and animals can even move between them (baby crocodiles, for example, won't have the size of adults that would allow them to keep warm just through being big).

All of these variations and issues belie the difficulty in figuring out the physiology of dinosaurs. There are, however, some important lines of evidence that can be brought to bear on the problem, to enable us to gain some idea of the metabolism and activity of the dinosaurs.

## Is walking tall the key?

One of the key early arguments for the possibility of high levels of activity in dinosaurs was the fact that they had an upright posture, just like birds and mammals, and unlike other reptiles and amphibians. The legs of dinosaurs are right under the body and they are geared for efficient and fast locomotion, with tyrannosaurs being an especially apt example of this, having long legs and running adaptations like long metatarsals. Inevitably, the situation is rather more complex. Crocodilians are actually capable of an upright posture (called the 'high walk'), even though they more normally sprawl like lizards, with the legs held sideways from the body. However, this may be a consequence of adapting to a semi-aquatic way of life, and some Mesozoic crocs and indeed other archosaurs were fully upright (so too were pterosaurs, for example). This suggests that both laterally positioned limbs, and ectothermy in modern crocs and their kin, may be secondarily acquired characteristics, rather than ancestral ones. The archosaurs as a whole may have been quite active, and if not necessarily endothermic, were perhaps moving towards being so.

In his well-known book *The Dinosaur Heresies*, palaeontologist Bob Bakker pointed to the fact that the fossil record suggested a predator-prey ratio of dinosaurs that was similar to that seen in modern mammals. For every carnivorous theropod we found, there would be perhaps two-dozen herbivores, exactly the situation we see in modern, mammalian-dominated ecosystems, and suggesting a high metabolism for them. The idea was quite novel and brilliant, but of course the problem here is that fossils relate to exceptional individuals, and are not necessarily representative of the actual population of the animals when they were alive. As we will see, one big issue with the fossil record is the lack of young dinosaurs, which probably made up a large percentage of the natural population and were the favoured prey of carnivores, so it's hard to know how much trust to put in this data. Given the tiny number of theropod fossils recovered, it's hard to be sure that what is revealed in the fossil record isn't any kind of statistical blip, as one extra find can throw out a low ratio to quite a degree, but overall it does at least point towards something like endothermy.

There is some evidence that a number of dinosaurs were active in places that were quite cold and were perhaps not suitable, or at least difficult, for ectotherms to survive in. Some dinosaurs were probably nocturnal, as demonstrated by their large eyes and unusual ears (like the small dinosaur *Troodon*), but more significantly, many lived at very high latitudes. While during much of the Mesozoic the climate was warm and dinosaurs had the benefits of a hot-house world to enjoy, in the Cretaceous those living in the northern parts of Canada and southern Australia would have seen plenty of snow and ice, and frozen winters. There is evidence to suggest that the dinosaurs were hanging around in these environments during the winter, so they could clearly tolerate cold temperatures that other reptiles could not.[19] Although this is the case for some lizards and amphibians that hibernate through cold spells, it does rather point towards a higher body temperature for the Arctic and Antarctic dinosaurs.

A great many theropods, including the tyrannosaurs, also had feathers. Not all of these were suitable insulators, but they were of a form that would provide insulation in *Dilong* and *Yutyrannus* (both of which, incidentally, herald from the Chinese Jehol fossil beds, which is thought to have been quite cold in winter), and would certainly have helped to keep in any body heat. *Yutyrannus* at adult size may have been large enough to exploit its mass to help it stay warm, but young animals (assuming they were similarly covered) and *Dilong* would not have been able to do this. As in moths, this argument is not wholly without exception, but at least it points to both endothermy and homeotermy. Of the few 'fluffy' ornithischians that had feather-like filaments, *Tianyulong* – another Jehol dinosaur, but this time an ornithischian – was positively heaving with fluff, which might well have helped to keep it warm.

By far the most convincing line of evidence in support of endothermy or at least homeothermy lies in the growth rates of dinosaurs. This subject is covered in more detail later, and it is enough to say here that their growth rates were utterly extraordinary, and commensurate or even superior to those of birds and mammals.[20] Dinosaurs, and in fact *Tyrannosaurus* in particular, grew *really* fast. This characteristic cannot be maintained unless an animal is a homeotherm, and baby dinosaurs would have been unable to exploit size to maintain a high temperature. They must have had a high metabolism for growth to have occurred at such a speed, and that can only mean that they were endotherms, at least when they were small.

It is also worth noting that birds are dinosaurs and homeothermic endothermy is near universal in them, and has certainly been inherited from ancestral birds, so it is not unreasonable to think that the Mesozoic forms also possessed this characteristic. Given how hard it is to tell apart most avian-like dinosaurs from true birds, and how many bird features there were in the theropods, it's also likely that endothermy was not so much a modern bird character but one that many theropods possessed. This doesn't necessarily mean that tyrannosaurs possessed it, but at least it raises the possibility that they did. Similarly, there are some good lines of evidence suggesting that pterosaurs were endotherms (for many of the same reasons: they had insulation, upright posture, fast growth and of course must have been pretty active to fly), and as noted above, perhaps some early archosaurs were, too. If that was indeed the case, endothermy would seem to be ancestral in dinosaurs, and there is a strong possibility that these animals were originally endothermic homeotherms.

## Heating up

Collectively, then, there are a number of reasons to consider many dinosaurs to have been both endotherms and homeotherms. In the case of those living in the warmer climates of the Mesozoic, homeothermy was probable even if they were ectothermic animals. Similarly, the large dinosaurs (especially the bigger sauropods and various ornithischians: animals that would be 5–10 tonnes or more) would have struggled *not* to be homeotherms. They were so huge that it would have been impossible for them to get rid of any heat generated by the simplest of activities, and must have held a fairly high temperature consistently. Indeed, if they were endotherms, overheating would have been a major issue at this size, yet they maintained a rapid growth rate when small. Might they have shifted their physiological processes, transitioning from endothermy to ectothermy as they grew?

This interesting aspect of growth is one that could also apply to the bigger tyrannosaurs, as they also changed size markedly as they grew (a newly hatched *Tyrannosaurus* might be only a metre or so long and weigh just a few kilos). The largest tyrannosaurines were, of course, huge and in the multi-tonne range, and this suggests that overheating at this size would have been a much bigger issue than getting warm. Interestingly, it has been suggested that feathers may have helped keep animals of this size cool. This sounds counterintuitive given all the arguments in favour of insulation for feathers, but it does make sense. While normally a densely packed layer of feathers (or indeed fur in mammals, and the fibres of pterosaurs and some ornithischians) would trap a layer of air and help keep the heat in close to the body, feathers could also be used to get rid of heat. If the feathers were more sparsely distributed across the body, they would do little to trap much air and thus keep in little heat (rather like most human body hair).

However, the bases of modern feathers have a small blood supply, which is essential for it to grow, but also potentially a way of eliminating heat. If these feathers were raised up and blood was pumped through the bases, this would effectively increase an animal's surface area and allow the heat to escape. Feathers would probably have been an important insulator in small and young tyrannosaurs, but could have had a very different purpose in large adults.

For most if not all the tyrannosaurs it is possible to make a pretty strong case for homeothermy and indeed endothermy. They were upright and active animals for sure, but they also had plenty of feathers (and ones that would seem suitable for insulation), were not very distant relatives of birds (and did have a number of other avian features), could be found in cold environments and grew extremely fast. It is thus most likely that collectively the tyrants were homeothermic at least, and possibly endothermic (though larger forms might have been ectothermic at adult size). In the day or night, in the far north or in the tropics, awake or asleep, a tyrannosaur would have been warm to the touch and capable of springing into action at a moment's notice. Moreover, it would have been capable of sustained activity and might have been able to run for quite some time if a situation required it to do so. It is impossible to know if tyrannosaurs were best suited to some kind of high-speed burst like that of a cheetah, or whether they used a continued pace like that of a wolf, but they would have been capable of something akin to a mammalian or avian output unlike the typical short bursts of crocodilians and other reptiles.

One critical relationship to metabolism is body size. Although there are inevitably exceptions, smaller animals generally have a higher basic metabolic rate than do larger ones, and as a result they rather burn through life. The axiom of live fast, die young very much applies to animals such as mice and small lizards, which typically live for only a few years, while ostriches and lions may live for a decade or two, and giants like elephants, whales and the biggest sharks and crocs can live to half a century or more. This implies that the large dinosaurs had potentially long life spans. However, even the large tyrants may not have lived much beyond their second decade, though the evidence is a little sparse. Few animals get to die of old age in the wild, and few large adult tyrannosaurs have been found. The suggested age of the largest individuals found is only around 20 or so years,[21] though the maximum (or even 'normal') age, for an ageing tyrannosaurine could have been rather higher. However, there is currently no evidence to suggest that they lived into their fifties or similar long periods that one might expect for mammalian multi-tonne animals, so they do appear to have been rather shorter lived than some modern and similarly sized animals, no matter what their metabolism.

## Size matters

Just how big were the tyrannosaurs and how do we know this? This is a rather contentious issue, but one that has major implications for the biology of the animals in question. Body size, and in particular mass, is perhaps the single biggest factor of an organism's life, influencing resource requirements, locomotory speed and efficiency, metabolism, growth rates and even behaviour. In short, this is something we really want to know.

The most often-quoted numbers for dinosaur sizes are total length values measured from the tip of the snout to the end of the tail. This is obviously a good start, and with a complete skeleton all you need is a tape measure – but it does come with several rather big issues. It's certainly useful to compare animals such as tyrannosaurs or at least the large theropods. However, oviraptorosaurs and therizinosaurs are built differently from most other dinosaurs, so a 5 metre-long tyrannosaur might be much heavier than a 5 metre-long oviraptorosaur. Comparisons between other dinosaurs are even more problematic: ceratopsians are not built like hadrosaurs, let alone sauropods or theropods, so again a simple length measurement has limited value. Moreover, many skeletons are so incomplete that any length is a bit of a guess. There are, of course, enough consistencies between taxa that it's quite reasonable to extrapolate between them: *Zhuchengtyrannus* is known from very few remains, but we do know that it's very close to *Tarbosaurus* and *Tyrannosaurus*, and those two are so similar that we wouldn't expect any major differences. Finally, using total length as a measurement involves including the tail, and that is a particular issue: even vaguely complete tails are rarely found for dinosaurs, and in cases where we do have them it is clear that they vary enormously even between closely related species and within species.[22] When measuring the length, it is therefore best to refer to the head, neck and body as a unit and exclude the tail. Even so, an estimate of the mass would be much better.

Mass, though, is if anything an even more complex issue: it can't be measured at all and has to be calculated or estimated in some way. Early researchers pretty much guessed how heavy dinosaurs may have been, and as a result few numbers had much in common with each other. Things have moved on since then and there are now a variety of methods of producing estimates of the mass of dinosaurs, with the popularity of tyrannosaurs ensuring that they have had more than their fair share of attention.

One relatively simple measure is the length and/or circumference of the femur. In almost all terrestrial animals – amphibians, reptiles, birds and mammals – of very different sizes and shapes (including both bipeds and quadrupeds), there is a fairly close correlation between both these measures of the femur and the mass of the animal to which it belongs. There's no particular reason to think that this relationship would not hold at larger sizes, so we can simply estimate the mass of tyrannosaurs by plugging them into the equation. More complicated measures involve trying to reconstruct the animal as a whole and working out the mass from there.

Early estimates of this type involved creating scale models of the dinosaur in question, dunking them in water to determine the volume, then scaling that by typical values of density for animals. Although we had almost whole skeletons of a number of dinosaurs, including *Tyrannosaurus*, when this method was first mooted, our understanding of dinosaur anatomy was limited, however it has improved considerably since then and exposed the flaws in such a simple method. We now know that tyrannosaurs had pneumatic bones and the air-sac system, and can adjust values for mass accordingly, and we also have a better knowledge of the sizes and shapes of the muscles. Newer models therefore run the gamut from relatively simple geometric models, to those based on laser scans of entire skeletons with individual groups of muscles restored, and different densities for different tissue types taken into account.

Many of the values that have been generated by such methods are actually quite close to each other, giving us some confidence in the results. No one value is ever going to be accurate, however, because there is typically quite a lot of variation between individuals of a single species, and even an individual organism can change a lot over a lifetime or even from year to year. Some animals, like bears, which can lay on a lot of fat for winter, may effectively double in weight from the start of spring to the end of autumn (then drop back down again in weight in the following year), and in these cases it would be equally correct, or indeed equally incorrect,

**Above:** *Tyrannosaurus rex* – the first to be discovered – on display at the Carnegie Museum.
**Below:** Three *Tyrannosaurus* specimens of different ages on display at the Los Angeles County Museum.

**Above:** A near-complete skull of the tyrannosaurid *Bistahieversor*.

**Right:** The flattened skull of the tyrannosauroid *Yutyrannus*.

**Right:** The well-preserved, but crushed, skull of the early tyrannosaur *Dilong*.

**Right:** The crested proceratosaurid *Guanlong*.
**Below:** The long-snouted *Qianzhousaurus*. This specimen lacks intact teeth.

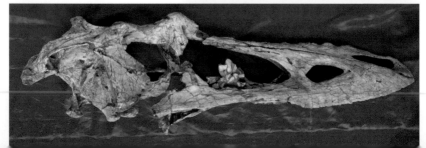

**Left:** Tyrannosaurs had feathers – these are preserved alongside the tail of *Dilong*.

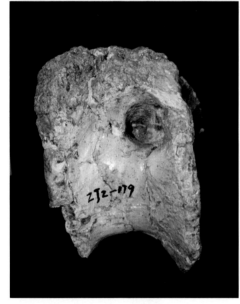

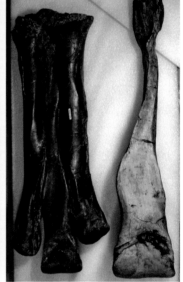

**Above:** A hole in a dorsal vertebra where an airsac would have invaded the bone.

**Above:** The three main metatarsals (left). On the right is a middle metatarsal – a very odd bone.

**Left:** Like birds, tyrannosaurs had a 'wishbone' – here is the furcula of *Albertosaurus*.

**Above:** Arguably the best tyrannosaur specimen in the world, a complete *Gorgosaurus*.
**Below:** Excavation of a juvenile *Tarbosaurus* maxilla, Mongolia.

**Above:** The humerus of a hadrosaur from Mongolia that shows feeding traces from the premaxillary teeth of *Tarbosaurus* – deep, parallel grooves in the bone.

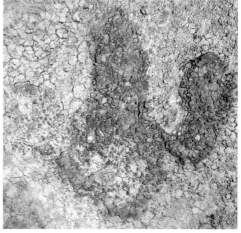

**Above:** A typical maxillary tooth of a tyrannosaur – here *Eotyrannus*.

**Above:** A rare fossil footprint referable to *Tyrannosaurus*.

**Above:** A giant coprolite attributed to *Tyrannosaurus*. This is full of bone fragments from a young hadrosaur.

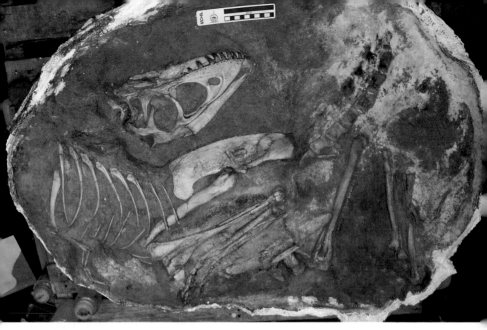

**Above:** Juvenile specimen of *Tarbosaurus* from Mongolia. This animal was only around a year old when it died.

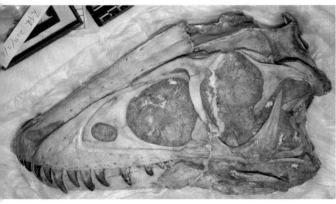

**Left:** Close-up of the juvenile *Tarbosaurus* skull.

**Below:** Very young and young skull restorations for *Tyrannosaurus* (to scale), based on scans of fossils and digital restoration.

**Above:** Digitally restoring parts of the head of *Tyrannosaurus*, including the eyes, brain, nasal cavity and sinuses, based on the elements of the skull and comparison with living archosaurs.

**Right:** Digital restoration of the muscles of the hindlimb of *Tyrannosaurus*.

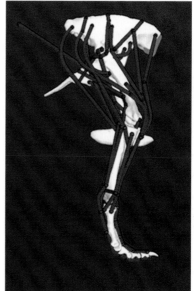

**Below:** Life reconstruction of *Tyrannosaurus rex.* By Scott Hartman.

A magnificent *Gorgosaurus,* on display in the Royal Tyrrell Museum, Alberta.

to go with either value, and an average of the two would still not reflect the reality. We can always work towards better mass estimates and refine more details as our knowledge increases, but these kinds of variation will always be there, so even a 'perfect' estimate would only be applicable until the animal has had a heavy meal, or has gone without food for a week or two.

The values that have been derived for the largest tyrannosaurs are really very big by the standards of living terrestrial animals: the larger specimens of *Tyrannosaurus* can measure around 7 tonnes or so.[23] The largest tyrannosaur specimen that is mostly complete is the famous 'Sue' at the Field Museum in Chicago, US. Measuring more than 12 metres long (with perhaps another metre or so of caudals missing), this is the biggest animal for which we can be confident of the size. However, Sue is not the largest tyrannosaur ever. For a start, there are very incomplete specimens that are larger, with various dorsal vertebrae that are bigger than Sue's largest vertebrae. Moreover, it is far from certain whether *Tyrannosaurus* was the largest tyrannosaur: one incomplete dorsal that is from the *Zhuchengtyrannus* quarry is at least comparable to the biggest in Sue. Not only that, but the specimen dubbed 'Scotty' seems overall to have been heavier than Sue. Despite being smaller in most linear dimensions, Scotty has both a longer and thicker femur, suggesting a more solidly built and heavier animal.

It must also be borne in mind that the odds of us ever finding the largest individuals are vanishingly remote. We have at best perhaps 30 or so decent specimens of the largest tyrannosaurines, and double that again in incomplete material that could be used to obtain a decent size estimate (a large dorsal vertebra, a femur or most of a skull is good for this purpose, a rib or partial foot less so). At any one moment in time at the end of the Cretaceous there may have been tens or hundreds of thousands of tyrannosaurs alive worldwide at any one moment, and over a few million years there would have been tens of millions or even billions, so the chance of finding a real giant in our sample of a hundred is tiny. However, giant tyrannosaurs must have existed. Every so often a huge member of a species rolls around that is quite a bit bigger than the average or even typical large examples; record sizes for all kinds of animal can be absolutely colossal. For example, a fairly typical male giraffe is around 5 metres tall, but the record is in excess of 6 metres, and a large male savannah elephant weighs around 5.5 tonnes, while the record is about double this. I'm therefore confident that for all the size of Sue and other apparently large tyrannosaurines, there were animals at one time or another that were considerably larger – hard though this might be to imagine. For anyone getting overly excited and thinking that this means we will actually find them, it sadly just won't happen. It should also be noted that the tyrannosaurs may well not have been the biggest terrestrial carnivores of all time, because the same is true for all the other giant carnivores that once roamed the Earth. Still, it is nice to think about giant tyrannosaurs and imagine just how much larger they could have been than the already quite mind-boggling sizes of the ones known to have existed.

The early tyrannosaurs were actually quite small, at only a couple of metres long, but the changes from small to large sizes were only one of a whole suite of modifications and evolutionary shifts that occurred over 100 million years of tyrannosaur history. These changes and what they meant for the lives of different species is the next subject tackled.

# CHAPTER ELEVEN

# Changes

We have identified the various tyrannosaur species, put them into an evolutionary context with a phylogeny, and described their anatomy and the probable functions associated with certain anatomical specialisations. Aspects of tyrannosaur evolution such as the reduction in the number of fingers and the decreasing number of teeth in the later forms have already been mentioned in brief. Here we look at such changes in more detail, and suggest what they might have meant for the tyrannosaur lineages at various times.

Starting at the very beginning, what was the ancestral tyrannosaur like? This animal must be hypothetical, because finding a representative of the true stem population that branched off from the other theropods is extremely unlikely. Early populations of ancestral animals are thought to have been small in terms of numbers of individuals, and generally short lived in evolutionary terms, so they are unlikely to appear in the fossil record. Moreover, cladistics does not allow us to identify such groups; if we ever did find such an animal and put it into a phylogeny, it would appear as the nearest relative of all other tyrannosaurs, but not as an ancestor. Part of the frustration palaeontologists have with the media lies in the fact that every new discovery is hailed as either an ancestor of something, or the missing link between two things, when it is neither (or more accurately we don't know whether it is or not). When considering what the ancestral tyranosaur may have been like, it must be a construct of our understanding of the lineages around it.

The ancestral tyrannosaur would have looked very much like the earliest forms we do know of – animals like *Guanlong*, *Stokesosaurus* and *Eotyrannus* (Fig. 12a) – but would also have been close to other early representatives of the groups nearest to the tyrannosaurs. These included the

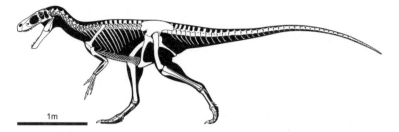

*Fig. 12a The early British tyrannosaur* Eotyrannus *that lacks many of the more specialised features associated with the later, and larger, tyrannosaurines.*

compsognathids, and perhaps either the ornithomimosaurs or the allosauroids (since it is not entirely clear if tyrannosaurs branched before or after the compsognathids). Actually, it may not matter too much, since we would expect such animals to be very similar to each other: they were, after all, *just* starting to separate from one another. Indeed, if you look at, say, *Dilong* alongside the Chinese compsognathid *Sinocalliopteryx*, there are almost no immediate or obvious differences between them at first or even second glance, and you could be forgiven for assuming that they are part of the same clade. This is because they are both early members of their respective groups and have yet to diverge much from the ancestral form.

Our tyrannosaur ancestor would have been small: perhaps only 2 metres or so in total length (half of which would be tail), and this is no big surprise. A small size is a common feature of early members of many evolutionary groups, since evolution tends to occur most rapidly in small animals (which have shorter generation times and higher populations than large ones), and extinction tends to finish off larger species first (giving the opportunity for survivors to diversify). What would distinguish our ancestor from other theropods on close inspection would be that it had the distinctive characteristics of the tyrannosaurs – or perhaps it might be some transitional or intermediate form between other theropods, and even the earliest of 'true' tyrannosauroids.

This ancestral animal would have had simple filamentous feathers. As previously noted, feathers or a version of them might just be ancestral for all theropods in any case, but given their presence in animals like *Dilong*, and all lineages that came later, they must have evolved before the tyrannosaurs split. However, it is the teeth that are critical here – the unusual tyrannosaur dentition, including the more robust premaxillary teeth, may have been a distinguishing feature in these very first animals. Their exact origin is a mystery, but their development was probably a fundamental part of basic evolution by natural selection: the shape and pattern of the teeth were modified in a new mutation within some small population or a few individuals. Variation is inherent in biology: however much you look like your parents or siblings, there are always some differences. New combinations of genes and/or alterations to them through mutation lead to individuals that are slightly different from their nearest relatives. Then some event (such as a drought, floods or the arrival of a new carnivore) puts pressure on a carnivore and a modification to the teeth gives it an advantage, leading to its spread and increase, or perhaps the change occurs in some isolated population, allowing it to spread rapidly. Selection pressures (such as competition from other predators or a lack of food) lead to the propagation of such a feature because individuals without it fare less well than those that have it. Such changes rarely occur in isolation and may be accompanied by new behaviour, and other parts of the anatomy might well adapt to the changing conditions.

A population in which such changes occur increasingly separates and diverges from its relatives (especially if it is isolated, or if some disaster has killed off its kin), and over time it is likely to become fully separated in an evolutionary sense. We regard such a population as a phylogenetically separate lineage, and can identify anatomical differences that make it different and distinctive from its relatives. The separated animals effectively stop interbreeding with the original population due to its extinction, geographic separation from it, some genetic difference that prevents interbreeding, or some behavioural modifications that result in the two groups of animals no longer recognising each other as potential mates.

We can observe the changes to the anatomy of these animals from the fossils, and correlate them with certain functions. We know that the presence of an arctometatarsus (pinched middle bone of the foot) and a stiffened ankle correlates well with efficient and even fast running, and the fact that this trait evolved multiple times in various theropod groups suggests that at some level it was evolutionarily 'easy' to acquire – although we don't quite know what prompted selection for this feature. Would chasing fleet-footed prey have driven selection for faster and more efficient carnivores, was a migration required to keep up with herds of prey where these dinosaurs lived, leading to selection for efficiency, or were other predators threatening them, giving an advantage to those that were faster?

These questions are hard to answer in part because we cannot identify ancestors, so we are forced to look at the changes in animals close to the separation of groups or in early members to determine what might have been happening when things changed. If we can combine this knowledge with information on other patterns (such as changes in climate, the evolution of new animal groups, other anatomical shifts and movement between continents), and our knowledge of dates and times of evolutionary branching events, we can start to piece things together.

## Getting a head

The changes we can most easily identify are ones associated with carnivory. Almost all features in organisms fulfil more than one function (horns on sheep, for example, are primarily used by males to fight one another for females, but may also act to fend off predators and have a small role in cooling the brain linked to blood flow through them), but most features would seem to have one primary function, or at least one that dominates their early evolution. When it comes to features intimately associated with carnivory, these are relatively unambiguous. True, the legs propel a carnivore whether it is hunting or not, but we can predict that the most critical phases of its life will be down to catching a meal, where every miss might be its last. This factor would thus have a stronger effect on the animal's survival than many other events. Changes to key features like the head, arms, legs and supporting structures (such as the neck for the head) may therefore be indicative in this respect.

The change in head size and shape was an obvious and major transition (Fig. 12b). Leaving aside the crests of the procertosaurids that rather affected their appearance, the skull

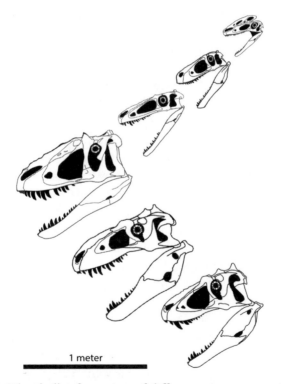

1 meter

*Fig. 12b The skulls of a variety of different tyrannosaurs. Although later members were characterised by their huge and robust heads, there were quite a few different forms with varied numbers of teeth. Shown top to bottom:* Guanlong, Eotyrannus, Alioramus, Yutyrannus, Bistahieversor & Lythronax

was essentially quite small in early forms. In *Guanlong* the skull appeared to be in proportion with the rest of the body: it wasn't some outsize monstrosity as in *Tyrannosaurus*, but would not have looked out of place on almost any other theropod. In the later tyrannosauroids, however, the head of *Yutyrannus* was larger overall (and in particular was rather tall), and started to look like a dominant part of the animal. This trait was similar in *Alioramus* and *Daspletosaurus*, while in *Tyrannosaurus* and *Tarbosaurus* the head really dominated. It didn't just get larger and taller (alioramins apart), but also became more robust. In the early members of the group, the struts and stanchions that make up the outline of the skull, and especially those around the fenestrae in the sides of the skull, were rather thin – there's a lot of space between them – but in the tyrannosaurines they were much more solid. Similarly, the dentary and mandible as a whole became deeper and more robust moving from the early to the late form, and they were especially exaggerated in the tyrannosaurines. Overall, the head had a lot more bone in it than before, and while this fact alone didn't make it stronger, it does suggest that these animals were developing an ever-more powerful bite and a skull capable of taking powerful forces.

This change came with others, most obviously in the teeth. As we move from the first tyrannosaurs to the later ones, two major patterns are in evidence: the number of teeth in the maxilla and dentary tend to reduce, so there are fewer teeth in the jaws, and those teeth increase in size and, in particular, are broad. In general, the teeth of carnivorous theropods are laterally compressed, so they are rather thin, and although they are certainly not 'blade-like', as is often stated, they are relatively narrow in cross-section. In the albertosaurines and especially the tyrannosaurines, the maxillary and dentary teeth are very fat, and if not actually circular in cross-section, they are certainly heading towards it. These two changes (thicker teeth, but fewer of them) would also seem to correlate with increasing bite power (as indeed would the shorter necks of later animals, which acted as a better support to the heavy head) and the changes wrought to the skull. This suggests that selection acted on all these features to alter how the animals used the head in biting, and gave them an increased ability to deliver more powerful bites.

Interestingly, the shapes of the skulls of the early tyranno-sauorids are rather like those of juvenile tyrannosaurids, suggesting that the head shape of the later animals was produced by changes in the patterns of growth as the animals got larger. This also suggests that the ecology of young tyrannosaurines may have been like that of adult tyrannosauroids,[24] since a similar shape would correlate with a similar style of biting.

### Reaching out, or not

Moving along the body, the arms show the opposite trend to the head, becoming smaller over time. Early tyrannosaurs had arms that were relatively long, and had three fingers, but moving into the tyrannosaurids especially, the arms reduced until they became stubby appendages in the tyrannosaurines. Moreover, although the unguals remained and appeared to have supported sharp claws, these were less curved than those of earlier forms, and their bases were reduced. The animals would thus have had less gripping power, with smaller ligament attachments. Combined with the relative bluntness of the claws, this feature implies that the arms saw less use than those of the earlier members of the tyrannosaur clade.* Once more this appears to have been linked to the overall importance of the head – after all, if the head is doing all the work when it comes to killing and feeding, the arms become less useful. Selection may act to reduce the arms because a large arm that is not being used is something of a dead weight: it costs resources to grow and lug around while not providing any benefits.

Interestingly, something similar happened in another group of large theropods. In the abelisaurs of the Cretaceous southern continents the arms were also reduced, while a more specialised and unusually short but deep skull evolved. These

*Despite all the gags about *T. rex's* tiny arms, *Tarbosaurus* actually had proportionally slightly smaller arms.

animals had still further reduced arms compared with those of the tyrannosaurs, and although they retained four fingers on each hand, these were tiny and either lacked claws or had very simple ones, while the humerus in particular was rather simple and almost degenerate. This, then, suggests that in both these groups the evolution that led to the change in the crania was linked to the reduced use of the arms and ultimately to them being altered as well.

Even the earliest tyrannosauroids had proportionally long legs compared with their body size, and this characteristic was maintained throughout the lineage. The last of the tyrannosaurines retained it and had a proportionally long femur and metatarsals, and this is something that separates them from all other non-avian theropods apart from the ornithomimosaurs.[25] Given the close relationship between them and the tyrannosaurs, perhaps it marked part of the original separation of these lineages. One of the key ancestral characters in tyrannosaurs was the elongation of the hindlimb parts, increasing the efficiency and perhaps the speed of the animals. It led to the ornithomimosaurs becoming highly specialised towards high speed, presumably in part to provide an escape option, since later forms appear to have been largely herbivorous. In the tyrannosaurs, however, it points perhaps to a new strategy relating to pursuing a meal, integrated with the changes to the head and arms – though obviously in their early history, small tyrannosaurs would also have been vulnerable to larger theropods and might well have needed to escape themselves.

At some point the metatarsals in tyrannosaurs were also modified, producing the characteristic 'pinched' middle bone, and the odd triangular shape of this part of the foot. This condition is not present in early tyrannosaurs, and is not fully seen until the tyrannosaurids appear. It is even apparently absent in the much later *Yutyrannus*, suggesting that it appeared relatively late in tyrannosaur evolution and only kicked in at larger sizes. Since this specialised foot is known to help reduce energy expenditure, perhaps this change was driven by the increasing size of the tyrannosaurs; the selective pressure in this case was the reduction of the effort of hauling around a multi-tonne body at relatively high speeds.

## Big is best

One other obvious trend is the pattern of increasing body size in the tyrannosaurs. The proceratosaurids and basal tyrannosauroids were small (under 5 metres in total length), the later tyrannosauroids were rather larger (around 8 metres), the albertosaurines and alioramines were larger again (10 metres), and the tyrannosaurines were the biggest of all. An increase in body size over time is not uncommon in various evolutionary lineages, in part because, due to the focus of extinctions on larger species, most clades start off at a small size. This means that to a degree the only way is up, though of course in the dinosaurs the increases in size were greater than in most vertebrate lineages, ultimately producing the largest terrestrial animals of all time in the biggest sauropods. Various dinosaur lineages got larger from small origins (the allosaurs, ceratopsians, titanosaurian sauropods and oviraptorosaurs, to name but a few). The exact reasons for these changes are not currently known, but I suspect that for the tyrannosaurs at least they probably represent a consistent evolutionary pattern, since they correlate with the ever-increasing bite power and adaptations towards this.

The trend is not absolute, however, since for a start there were animals such as *Tarbosaurus* living alongside smaller forms like *Alioramus*. More notably, *Nanuqsaurus* in the Canadian north was far smaller than we would immediately predict it to be judging by the size of its relatives. While still sizeable (estimated at around 7 metres in total length), by any comparison with other tyrannosaurines, it was small – arguably even a dwarf. What caused this variation in size increase? Moreover, the reduction in size is a double surprise, for *Nanuqsaurus* lived in the far north and we might expect an evolutionary phenomenon called Bergmann's Rule to kick in here. This is a pattern whereby animals that live in colder, more northerly latitudes tend to be larger than those that live towards the warmer south. In northerly latitudes, a larger body size helps animals stay warm (they have a smaller surface area to volume ratio), so northern species, or populations of species with broad distributions, tend to be larger in higher latitudes than in lower ones. *Nanuqsaurus* is both a tyrannosaurine and lived high up, so why was it so small? The probable reason for this is that it was isolated in a relatively small patch of land, and with limited resources available over a small area, the population would have been small and individuals would be selected to be smaller. This is the Island Rule, and here it seems to have overruled the expected pattern and led to the eventual evolution of a small tyrannosaurine.[26]

This brings up a point about the limitations of evolution. Clearly body size cannot continue to increase forever: fundamental limitations such as the physical properties of bone and muscle, and the limits of physiology to get in enough oxygen or transmit neural signals, eventually prevent a species from getting any bigger. In the case of *Nanuqsaurus*, a conflicting selective pressure was involved: a large size brings the advantage of a greater retention of heat, but this is balanced out and even exceeded by the limitations of living on an island and reduced resources. There are also the inherent problems involved in changing an existing form: evolution can only work with what is available, and when this is a derived tyrannosaur with small arms and only two fingers, a sudden shift in selective pressure that favours, say, delicate manipulation of food by the hands, is going to be difficult. In such cases specialisation may occur elsewhere in the body, involving an alternative solution to the problem such as a move into a different ecological niche, or even the extinction of the species. Lost features can be regained, but if they have been absent for a long period the genes for them may have been lost too, which will make it much less likely that they will return without a whole host of mutations kicking in together. Hence we see effects such as the co-option of features: they may be specialised for multiple functions, or involve a compromise between different selective pressures. The result may be amazing developments like the tusks of elephants, which can be used in combat with other elephants or to fight off predators, dig for water, strip bark from trees and more; no one selection pressure truly overrules the others, although sexual selection and intraspecific combat seem to dominate when it comes to tusks at least.

There are always such wonderful exceptions and deviations in biology. Given enough time and diversity, it's almost inevitable that there will be an evolutionary exception nestled somewhere in a clade; examples include giant flightless birds, hairless mammals, legless lizards, giant tortoises, and frogs that develop from tadpoles to adults. Even taking these out of the equation, there are some clear evolutionary patterns for the tyrannosaurs, and tracking these gives a great impression of how they changed and what the changes meant for their lives. In the next section we move to the tyrannosaurs' behaviour and ecology: how they fitted into their ecosystems, and their interactions with their own species and others.

# PART THREE
# ECOLOGY

# CHAPTER TWELVE

# Reproduction and Growth

There are currently no fossil eggs that can be confidently assigned to any tyrannosaurs – and no nests, hatchlings or embryos – so our knowledge about their reproductive biology is rather limited. There is hope, however: parts of Mongolia that have yielded fossils of *Tarbosaurus*, *Alioramus* and *Raptorex* are also known for huge numbers of fossilised dinosaur eggs of numerous species, a number of which do contain embryos, and there are also several fossils of therapod parents lying across nests.[1] As a result, it's far from implausible that eggs and even whole nests will turn up for some tyrannosaurs at some point in the not too distant future.

Despite the overall lack of data, there are some things that we can be reasonably confident about. For example, all archosaurs that we know of laid hard-shelled eggs. Although various lizards and snakes have evolved to give birth to live young (as have a number of marine reptiles), none of the crocodilians, birds, pterosaurs or any non-avian dinosaurs ever did so; it is therefore plausible to infer that the tyrannosaurs also laid eggs, and that these had hard shells.

A common pattern in dinosaurs and crocodilians (less prevalent in birds) is that they lay large numbers of eggs. Most intact dinosaur nests that have been recovered in fact contain large numbers of eggs – sometimes more than 50. Again, despite the lack of direct tyrannosaur data, this pattern occurs in numerous dinosaur species across the dinosaurian clade, and – although the data is concentrated in the hadrosaurs and ceratopsians – there is evidence that plenty of theropod representatives also laid eggs, including those relatively close to the tyrannosaurs, such as oviraptorosaurs and therizinosaurs. It's pretty unlikely that tyrannosaurs were doing something massively different from all other dinosaurs that we know of, or indeed from most living archosaurs, even if a few birds took a different route (kiwis, for example, lay just one, truly colossal egg).

One obvious complexity in interpreting this data is that some birds, at least, have a system where females may lay eggs in multiple nests that are brooded by males. Ostriches do this, so an ostrich nest that contains dozens of eggs may have had numerous females contributing just a few eggs each to it. If some dinosaurs did this, then the very high numbers of eggs we see in nests wouldn't provide an accurate representation of the fecundity of a typical female, but rather of the output of many. Two lines of evidence stand against this, however, and support the idea that dinosaurs were typically highly fecund.

First of all, while birds such as ostriches may contribute only a few eggs to a nest, they will typically do so to multiple nests, so the overall output is still high per female. Secondly, dinosaur eggs are typically small compared with the size of adult dinosaurs. There is a general pattern in animals when it comes to reproductive strategies: they either produce a few large offspring with considerable parental investment per baby (often exaggerated still further with care of the eggs and/or babies), or they produce large numbers of offspring with little investment in each. Contrast the strategy of rats, which might have a dozen babies every few weeks and wean them quickly before starting to reproduce again, with elephants that have one baby every few years, then spend years more rearing it and looking after it. Such behaviour can reach extremes in animals that produce huge numbers of offspring, and this is especially true of egg layers: just look at the numbers of eggs produced by some frogs and fish, for example.

Based on the small sizes of dinosaur eggs (even the very largest sauropods that clocked in at tens of tons in mass had eggs smaller in diameter than a typical football), it seems that the majority were taking the 'many-and-few' approach. In fact, they may not have had a lot of choice: there's a limit to how large a hard-shelled egg can be. A very large egg would need an extra-thick shell to provide the necessary support so as not to collapse under its own weight. Dinosaurs could have produced thicker shells, but doing so would have killed the developing embryo inside. Although they may look solid, reptile and bird eggs are covered in tiny pores, which allow oxygen into the egg for the baby to metabolise and waste carbon dioxide to get out. To make a giant egg that was, say, a couple of feet across (which one might expect for the huge dinosaurs, including the largest tyrannosaurines), the shell would need to be so thick that it would prevent gas exchange between the developing embryo and the external environment and the embryo would suffocate. Large dinosaurs therefore simply could not produce very large eggs commensurate with their size (in the way that a whale or rhino can have a very large baby), so would have had to go down the route of small eggs and, by extension, larger numbers of them. Some smaller dinosaurs do seem to have potentially gone for lower numbers of larger eggs, but this would not have been an option for the larger tyrannosaurs.

For most dinosaurs around the world there would have been an optimum breeding season when food was most plentiful and the weather was at its best, and reproduction would have been limited to this time. The best conditions for growth produce the maximum amount of plants, and by extension seeds, leaves and fruits. This is when herbivores of all kinds will want their offspring to appear, to take maximum advantage of the available food and lowered competition, and in turn when carnivores will want to take advantage of the glut of new babies. In the higher latitudes, of course, this spring period would have been very limited, and there would have been a real scramble for the animals to reproduce in the narrow time window available. Closer to the equator seasons would have been less of an issue, with food generally available all year round, but there are slight shifts in the seasons even in these areas, and there still tends to be a breeding season for equatorial plants and animals as a result. Some species do breed all year round, and the output in others is fast so that they can get through several rounds of reproduction in a relatively short space of time.

Dinosaurs may have produced multiple clutches of eggs as do some birds and crocodilians, so just because we find a typical nest for a given species having, say, 30 eggs in it, that doesn't mean this was a typical annual output for this animal. If they laid two or even more clutches a year, they could have been producing more than 100 eggs annually. This may seem extreme, but in fact it's quite plausible. Many birds, for example, lay small numbers of eggs because there is a maximum amount of chicks they can care for, rather than a maximum of eggs they can lay: the production of eggs is not a limiting factor. Animals that are not that healthy or can only secure a small territory lay fewer eggs than better positioned parents: they wouldn't be able to feed too many offspring, so simply cut down on the number of eggs they lay. Blue tits can produce dozens of eggs over a few weeks if their clutches are destroyed or removed, and added up they reach something close to the mass of the actual parent. Even hundreds of eggs over a few months could thus be a plausible number for some larger dinosaurs to have produced.

## A nested development

The dinosaur nests that we know of were relatively simple affairs, and not the complex interleaved vegetative structures of most modern birds. Roughly circular scrapes were excavated in the ground, and these may have been lined with loose leaves or similar to provide some protection, moisture and heat from decomposition. In short, the nests were generally more like the nests of alligators and crocodiles than those of most birds. However, some later theropods did sit on their nests as birds do,[2] and they may have used their feathers to help protect the eggs or newly hatched baby dinosaurs, and of course to help keep them warm.

Dinosaurs, including tyrannosaurs, probably exhibited some form of parental care over both the eggs and the newly hatched young. A few birds have all but eliminated parental care and look after the eggs in the nest, but not the hatchlings, which fend for themselves as soon as they hatch. However, this approach is very unusual, and all crocodilians and the rest of the birds look after both the eggs and the newly hatched youngsters. They may do this for as little as a few weeks, or a year or more, but there is generally some care from at least one parent (typically, but not exclusively, the female), and often both. There are a number of dinosaur fossils that appear to include preserved young juveniles in the company of adults, which may represent parents living with their offspring, and the interpretation is reasonable as this is what we would expect.[3] Still, given the lack of available data, this is about as far as we can go. All of this is probably true for tyrannosaurs, but the exact degree and extent of any parental care is hard to determine.

Naturally, passing on your genes and raising the next generation is a key part of the life cycle of any organism, so it's not surprising that animals attempt to maximise their reproductive potential. Part of this comes down to ensuring that they have a good mate: they don't want to be stuck with a partner who is sickly or a poor forager. Animals go to great lengths to ensure that they mate with an appropriate partner, and as explained below this can lead to some interesting evolutionary effects. It also means that they need to adapt their reproduction strategy according to the prevailing conditions. As noted above, animals typically go for either larger numbers of offspring with little investment in each, or few with considerable investment, but there are further subtleties to this and interesting possibilities for dinosaurs.

With egg size limited, dinosaurs produced more eggs for their size than might be expected, but if they were investing heavily in their offspring they may not have laid as many eggs as they could. Looking after babies can represent a huge amount of work: just look at how manic small birds are when their eggs hatch, with parents returning to the nest every few minutes to cram food down the gullets of their offspring. They also have to keep the nest clean, keep away parasites and fight off predators and other parents encroaching on their territory; it's no surprise that this can take its toll and leave them exhausted and vulnerable.

That said, even with large numbers of hatchlings, parental care would not be too onerous given the size of the babies; as in crocodiles and unlike in birds, a newly hatched tyrannosaur would have been tiny compared with the adult, and would not have required huge amounts of food, for example. A small bird may have only half-a-dozen chicks in the nest, but within weeks each one might be half the size of a parent bird and still reliant on the parents for food, so they would need to feed the equivalent of three or four adults each for days at a time. However, due to the smaller starting size and considerably longer growth period of tyrannosaurs, even after a year the youngsters might still be a small percentage of the adult mass. Coupled with the fact that quite a few babies would probably have died in the first few days or weeks (infant mortality in all species is huge, and it's been estimated that 60 per cent of young tyrannosaurs would not have made it past their first year),[4] it would probably have been better to produce a few more eggs – and hope for a few more to survive – than to hang back to avoid starving the existing brood. For something like an adult tyrannosaurine, a single typical prey item might have been enough to feed a whole nest full of month-old babies for several days, so the immediate care wouldn't have been a major burden the way it is for small birds.

Leaving aside some of the more uncertain issues, one thing we do know is that young dinosaurs in general grew quickly, and in fact *Tyrannosaurus* in particular grew astonishingly fast. Rather like trees, dinosaurs and various other vertebrates lay down something akin to a growth ring in their bones, so their ages can be determined with some accuracy. As they grow, new bone is laid down in ever-increasing layers on top of the last. In poor conditions (in winter) growth is slowed, so the layer added is thicker and appears darker, but in the good times the layers space out. So a simple counting of the darker rings reveals the number of winters, and therefore years, an animal has lived. We can obviously see how large the animals were, so with a number of specimens of different sizes we can chart their age versus their size, and see how quickly they grew. The ring-counting method isn't 100 per cent reliable – rings can be lost during growth and are not always laid down clearly, so some may be older than they appear – but even so, some of the rates recorded are quite extraordinary.

A hatchling of a large tyrannosaurine might be around only a metre in total length, and weigh just a kilo or two: something about the size of a large house cat or small dog. However, by the age of two it might be in the realm of 30 kg, and by 15 nearer to 2 tonnes. By the age of 20 or so, it might be hitting a fully adult size of say 10–12 metres long and weigh more than 5 tonnes.[5] This would imply a somewhat extreme rate of growth, apparently faster than that of other dinosaurs, including most other tyrannosaurs – compare *Tyrannosaurus* with *Daspletosaurus* and *Albertosaurus* in Fig. 13 – but all of these animals clearly grew rapidly.

Quite what facilitated such a rapid growth is not entirely clear. Certainly, the apparently high metabolism of these

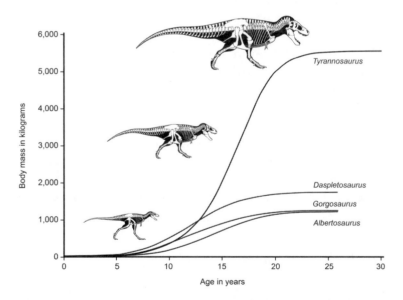

*Fig. 13 Large tyrannosaurs grew very quickly from just a few kilos as hatchlings to multi-tonne adults in under two decades.* Tyrannosaurus *seems to have grown quicker even than its other near relatives from North America.*

animals would have helped (if you have a higher metabolism, things happen faster, provided of course you are getting enough to eat), and if this increased as they got larger (or became more efficient, so more effort could be directed towards growth), then this would have accelerated matters as time went on.

Importantly, young animals may have benefited from the attention of their parents, and this is perhaps a strong argument for a heavy dose of parental care in the first year(s) of life of young tyrannosaurs. If food was brought to the young, this would have provided them with nutrition to grow, but critically it also would have saved them the huge amount of effort required to forage and find that food, and the energy and time wasted in avoiding predators that preyed on small and vulnerable young dinosaurs.

A comparison to baby birds here is instructive: they grow very rapidly, having typically little to do but sit in the nest and be fed. Although they often hatch at a size that is a considerable percentage of the adults' (thanks to those proportionally larger eggs), they can go from hatchlings to fully fledged, adult-sized animals in a matter of months or even weeks, so rapid growth can certainly be boosted by the parents and may point to some care in the tyrannosaurs. Mammals can grow similarly quickly, and again typically get a lot of care from their parents and are supplied with milk, while lizards and crocodiles – despite some occasional help or protection from their parents – do need to put their energy into finding and catching food, so perhaps inevitably grow more slowly (though of course their typically lower physiology also hinders growth).

## Ch–ch–ch–ch–changes

As animals grow they naturally change shape. A characteristic of baby tetrapods is that they tend to have large heads with large eyes, and large feet. Features we think of as endearing in puppies and kittens are really universals, and seen in humans as much as they are in dinosaurs. We have rather few fossil baby dinosaurs overall: they are rare in the fossil record for a number of reasons, being small and hard to find, and more vulnerable than adults to being eaten when alive, which precludes them from being fossilised. However, there are just a few fossils of immature tyrannosaurs from a variety of taxa. Many are actually close to adult size, but relatively young juveniles of *Guanlong*, *Tyrannosaurus* and especially *Tarbosaurus* are known, while the only known specimen of *Raptorex* is far from an adult animal.

We can trace changes during growth across a single species, and here *Tarbosaurus* is perhaps the best known. A recently described very young juvenile specimen was found in Mongolia. It was as little as two years old, and measures only 2 metres or so in total length, which includes a fair amount of tail.[6] We then have some nice adolescent material, and indeed a number of fossils of adults. While there is currently no decent fossil of a middle-sized juvenile for *Tarbosaurus*, there is a specimen like this for *Tyrannosaurus*, and given how closely related these two genera are, that will do as a suitable proxy.

As discussed earlier, through time there were major changes across the various tyrannosaur clades, and to a degree these can also be seen in the development of a single animal. Looking at the juvenile specimen of *Tyrannosaurus* known as Jane (of which more later) and comparing this to adult animals, we can see how things change. The skull gets larger compared to the overall size of the animal (a reversal of the normal trend in vertebrates) and also more robust, and the number of teeth appears to drop between juveniles and adults in several tyrannosaurs, though oddly not in *Tarbosaurus*.[7] The horns over the eyes also enlarge, further changing the appearance of the head. The limbs remain long but adults are not as 'leggy' as younger animals, while the torso is much larger and the adult animal is overall more robust.

This is a good illustration of the shift from a young animal to an adult one, but it obviously begs the question: if they changed so much when they grew, how would one tell an adult from a juvenile dinosaur? Size alone is going to be a pretty poor criterion since, for example, lion cubs are not too dissimilar from an adult house cat in terms of size, but clearly they are very differently sized animals when fully grown. Tooth wear is often referred to in mammals to figure out how old they are, but the regular replacement of teeth in dinosaurs rules this out. Instead, we can look at three key areas to help determine the age of a dinosaur.

First of all, there are the growth rings in bones referred to earlier. Growth might be extremely rapid, but an animal that is only a year or two old is highly unlikely to be an adult. Secondly, we can look at those classic proportions: if an animal has especially large feet, or a huge head with massive eyes, then it is probably a juvenile, and indeed a young one at that, because such a pattern is near universal in vertebrates. Caution is required here, though, since some animals retain relatively juvenile characteristics well into adulthood (some breeds of dog have huge paws even when fully grown, even though this is normally a feature of puppies). Finally, we can look at the fusion of the skeleton.

You may know that when they are born, humans have heads that are rather 'squishy': the various plates of bones that make up the skull are not yet fused together and can move around. It takes quite some time for the bones to fully fuse together and produce the single unit that we are used to thinking of as the human skull. If you take a look at a skull of an adult, you can still see the little zigzag lines between the fused bones, called sutures. Similarly, the pelvis is made up of three pairs of bones that fuse into a single big unit, and in particular each vertebra of the spine is actually two parts melded together (the centrum at the bottom and neural arch above). Dinosaurs are similarly built, with the bones fusing slowly as they get larger and closer to adult sizes. Critically, though, this happens in a relatively consistent pattern, so that with a decent amount of a skeleton we can have a go at assessing the status of an animal as anything from a hatchling through to an adult.

The two parts of each vertebra (the centrum and neural arch) slowly start to fuse together as an animal grows, until the suture is barely visible and eventually disappears entirely and can no longer be seen in adults. This also typically happens at the rear first and moves forwards, so the tail vertebrae may be fully fused, while those in the middle of the dorsals are not quite completely fused, and those in the neck may be quite a way from complete fusion. Add to this the state of the pelvis and skull, and we can gain a good idea of the relative age of a given tyrannosaur. For example, the single specimen we currently have of *Raptorex* has rather open sutures throughout its vertebrae, so clearly had a fair bit of growing still to do.

A fully grown adult tyrannosaur with fully sutured bones would be considered 'osteologically mature', but this does not mean that it would have finished growing. It is still not clear if dinosaurs continued to grow throughout their lives (as do crocodiles, lizards and various other vertebrates), or if they eventually stopped growing once they hit full size, as do birds and mammals. Certainly, when they got close to adult size growth slowed down greatly in dinosaurs, but it may have continued, albeit very slowly, until they died. Interestingly, dinosaurs were probably sexually mature well before they were osteologically mature. This is actually the case in plenty of animals, but is generally unusual in mammals and birds, although humans are a notable exception (we are typically sexually mature as young teenagers, but may still be growing into our early twenties).

As noted earlier, female dinosaurs exhibit medullary bone when they are laying eggs, and this has been found in a variety of dinosaur species. It has also turned up several times in animals that are not osteologically mature, indicating that they were effectively having children while still 'teenagers'. This is actually quite a reasonable reproductive strategy: most dinosaurs died young, and to have any chance of passing on their genes to the next generation, reproducing at the first opportunity may have been a good idea. Medullary bone has been reported for at least one tyrannosaur: a small but adult *Tyrannosaurus* specimen from Montana, in the US, which shows the special layers of medullary bone in the femur.[8]

Despite claims to the contrary, there are no apparent consistent differences between specimens of any tyrant (though especially *Tyrannosaurus*) that denote males and females. There is a commonly held misconception that females were bigger than males, but since we can't reliably tell one from the other in the vast majority of specimens, there's no reason to think that this was actually the case. In some animal species, females are indeed the larger individual on average (tigers, and many spiders and frogs, are a few examples), and this may correspond with additional investment of females in their offspring, relative to males. The males may be investing in a different way, however, by competing for females, which is why males are often larger than females.

Animals compete directly with other members of their species at various times in their lives. They may fight over food or water, over territory, to be the dominant member of a herd or for the right to mate. Any of these factors may be critical to the short- or long-term survival of an individual or its line, and may occur once in a lifetime or almost daily, depending on the situation. Interactions can be highly aggressive but without necessarily dissolving into violence. Watch two dogs that are gearing up for a fight and there is likely to be lots of snarling and growling, circling, raising of hackles and the like before anything actually starts. They are sizing each other up and somewhat ritualistically advertising both their size and dominance with vocal and visual signals. More often than not, one will back down before any serious fight has begun. Some things are worth fighting for, even to the death, but there's little point in getting involved if you are sure you are going to loose, so it is in the interest of both animals to see how it is likely to go before risking an injury or worse in a serious fight.

## It pays to advertise

One way in which animals compete is by using advertising signals. Features such as a big crest, a set of horns or large teeth can all indicate the relative health and overall status of an individual. The bigger the signal, the more effort and energy an animal has devoted to growing it and hauling it around, and therefore the fitter and stronger it is likely to be. Bright colours in an animal indicate that it must be in good health and also show that it can avoid the attentions of a predator despite being conspicuous, or if it is a carnivore itself, that it can hunt successfully despite risking giving away its position to potential prey. Such signals are useful in battles for status and resources, but can also be used to lure in potential mates.

Sexual selection is the branch of evolution that relates to changes which promote reproductive success: having more or better offspring. Classic aspects of sexual selection are the very fact that we have males and females at all (as there are different investments to producing eggs and sperm, and different degrees of certainty of parentage), and signals like the feathers of birds-of-paradise, or weapons like the antlers of deer. Such signals can take on multiple roles: they can be put to good use to fight off threats, as well as being adverts to the opposite sex or even general signals of social status. At least some dinosaurs fought each other (the horns of *Triceratops* are known to have been used to battle other members of the same genus, for example, as seen in dents and injuries on the skulls of a number of specimens), and some of the huge range of plates, crests, horns, fangs and frills on various dinosaurs were probably used as social or sexual dominance signals.[9] Tyrannosaurs were no exception in this respect.

The most obvious features are the elongate crests on the heads of the proceratosaurids *Proceratosaurus* and *Guanlong.* These are large compared with the heads that bear them, and comparable to the crests of many other dinosaurs in that sense, but unusual for theropods. Carnivorous theropods in general seem to have typically had smaller crests than those of herbivorous forms, and this makes ecological sense. Herbivores are present in large numbers and spend large parts of their lives foraging for food. They may be out in the open doing so, perhaps often in a herd, so a large visual signal that might give away their position is not much of a risk: it's generally not hard to find them as it is. In contrast, a hunting theropod with a large crest might be spotted more easily by potential prey than one without it. Sure, you want to advertise to mates and rivals, but such a feature is of little use if you are starving to death because your prey can see you coming, so it was perhaps limited by conflicting pressures of advertising versus being able to eat.

Interestingly, proceratosaurid crests are hollow and full of holes that look as though they are pneumatic, and were likely connected to the sinuses. Later tyrannosaurs didn't have anything like as large a set of crests, but most species did still have bosses over the eyes. These were especially large in *Alioramus*, but were also quite clearly present in *Yutyrannus* and other tyrannosaurs. Both these hornlets and the rugose patches of bone along the snout were probably larger in appearance with the addition of keratin, but their overall size was still relatively small compared to the size of the head and the skull compared to things like *Triceratops* or the crested hadrosaurs.

These various features were used for some form of signalling between animals; they certainly would not have had any mechanical function like acting as a point of attachment for muscles. One might think that only males would have had such features, as is common in displaying animals (like peacocks, elks and lions), but this was not always the case. Mutual sexual selection can occur where both males and females are in competition with their respective sexes for the best mates available, so both may have crests to signal their quality. This can also overlap with social/dominance signals, so the fact that all individuals have horns is not evidence against sexual selection. A growing number of animals is known to be under mutual sexual selection, and the increasing numbers of examples among birds points to this being a feature in their extinct relatives as well.

As described later, things certainly did get heated between tyrannosaurs at various times, and there were fights – perhaps even ritual battles – between individuals. This may well have included fights over mates, or even between potential mates (in some species males are effectively forced to overpower females before they can mate as a demonstration of their health), though we cannot confirm this.

As to actual mating, well, it certainly occurred. Males would have needed to mount females and transfer sperm, but beyond this we are edging into conjecture and speculation. Still, some things are more likely than others and there are some fairly obvious issues to get around. Number one is the fact that tyrannosaurs (and indeed many dinosaurs) are both large and of an awkward shape, and getting the two relevant parts together would not have been simple. Palaeontologists have tried to work out quite how various dinosaurs would have mated. The trickiest to figure out appear to be the armoured and spiky stegosaurs, with the giant sauropods a close second, but things are not simple for the tyrannosaurs, either (especially large ones), with the muscles around the tail getting in the way, in particular.

They could not easily get face to face, and that leaves a mating posture similar to that adopted by most animals (including extant archosaurs) of the male mounting the female from behind, with her either lying down or standing. It's actually been suggested that the small hands and arms were retained in the later tyrannosaurs for holding females in place when mating – though this hardly explains why the females also kept them. This would still leave a distance to be bridged, and here the most obvious solution, as noted before (for tyrants and plenty of other awkwardly shaped dinosaurs), would have been the evolution of a delightfully termed 'intromittent organ'. Then, with the eggs fertilised inside the female, the egg shells could be formed, the eggs would be laid, and a few weeks later the first of the next generation would begin to hatch and start snapping at the local fauna, on their way to becoming the latest great carnivores of the Mesozoic.

# CHAPTER THIRTEEN

# Prey

The world of the tyrannosaurs was a complex and changing one. Between their origins in the Jurassic and extinction at the end of the Cretaceous, dramatic changes occurred on the planet, including the break-up and movement of the continents, major fluctuations in the climate, and the development of flora and fauna. In the Jurassic, the first flowering plants appeared and began to spread; the giant cycads, ferns and horsetails that were ubiquitous in the early part of the Mesozoic eventually gave way to more modern trees and shrubs, including familiar ones like magnolias and ginkgos. The Cretaceous saw the first appearance of the grasses, which provided a totally new ground cover. Such changes to the plant life affected what and how the herbivores ate, and hence led to evolutionary shifts in these animals. They would also have altered the spacing and size of cover for predatory animals.

Although in the course of time the theropods changed somewhat in form across various lineages, the carnivores were as a whole relatively conservative in this respect. All were bipeds, none ever evolved any kind of bony armour or plates, and while a few (including early tyrannosaurs) did have crests on their heads of a sort, these were relatively small. By contrast, in the ornithischians and sauropods there was a colossal range of sizes and shapes in body plans: for instance, some walked on two legs, or four, or could switch between the two, others had huge necks or massive tails, and there were animals with armour, spikes, plates, hefty tail clubs of bone, horns and giant head crests of great variety. Even the herbivorous theropods were often rather odd compared with their meat-eating cousins, and while they might have shed their sharp teeth and claws, they also had some other interesting evolutionary patterns that made them quite different from their cousins.

Dinosaurs were ancestrally carnivorous. The dinosauromorphs and their nearest relatives (such as early pterosaurs and crocodilians) were all carnivores, and the various herbivores diverged from their original carnivorous diet, while the theropods kept true to their origins. Vegetarianism might seem like a cheap meal-ticket for an animal: there are, after all, huge amounts of plant material available, and plants don't tend to run away when you try to eat them. However, plants do have all manner of defences against being eaten, such as thorns and toxic leaves, while the leaves and stems of grasses contain silica, which is very wearing on teeth and hard to digest. Moreover, plants are not very nutritious compared with meat, and animals need to eat a lot of leaves (and even more of things like bark and stems) to get much from them, so they have to either eat a huge amount, or digest their food for a long time, or both. All of this has to be done against a background of competing with other herbivores for food, and of course the threat of predators. Herbivory is nowhere near as easy as it appears, and just as carnivores develop adaptations to their ways of life and to separate them from their competitors, so too do vegetarians.

It is important to figure out what the tyrannosaurs were trying to eat, and which species were (or were not) on the menu, and how and where these animals lived. The meat of a dead animal is pretty much uniform, but it is a very different matter to try and lunch on an ankylosaur covered in bony plates and spikes, and wielding a club on the tail, to tackling an ornithomimid with not a scrap of armour but capable of very high speeds. So here we take a detour away from the tyrants and look at the animals that were on their menu, the implications this had for the fauna as a whole and, when we return to tyrannosaurs, on their own ecology and behaviour.

## Heterodontosaurs

This group of ornithischians was one of the first to branch off and diversify and was primarily part of early dinosaurian faunas, although a number of the animals survived through the Jurassic and even into the Cretaceous. All of them were relatively small (under 2 metres in total length), and were primarily bipedal herbivores. Dinosaurs were, on the whole, actually quite large animals as adults,[10] and there are few very small taxa that are not maniraptoran, some of the more bird-like theropods, so the heterodontosaurs are one of the few groups so far known to contain only small species. They were probably primarily herbivores – although it has been suggested that they might have also taken insects or similar small prey on occasion.

Heterodontosaurs had a small beak at the front of the snout and rows of teeth behind, though their main feature was a large and fang-like tooth on either side of the mouth that extended from the lower jaw. This gave the group its name (which roughly means 'different-toothed reptile'), and has resulted in heterodontosaur skulls having something of the appearance of a pig or muntjac deer, with a rather obvious giant tooth in the jaw. Heterodontosaurs also had a kind of cavity built into the maxilla that the fang slotted into, so it fitted rather neatly when the mouth was closed. The function of these large teeth is not clear: they may have been used to tackle animal prey, in threat displays or in fights between animals as are the similarly shaped teeth of baboons.

Moving down the body, heterodontosaurs had some large claws on the first digits of the hands, and these may have been used to take insect prey to supplement an otherwise herbivorous diet. Unlike in almost all other ornithischians, the tail did not have a series of ossified tendons running along its length, which would have strengthened and supported it, so the tail of a heterodontosaur would have been more flexible than the tails of most ornithischians. One important aspect of the appearance of this clade is that the heterodontosaurs are one of the limited set of non-theropodan dinosaurs known to have exhibited some kind of filamentous covering. Several specimens of the genus *Tianyulong* from the Early Cretaceous of China have been found preserved with a thick coating of long, hair-like filaments running along the back from almost the snout to the tip of the tail,[11] which would have made the living animals look like some kind of overly fluffy little beasts. These animals seem to have favoured arid or semi-arid environments, though *Tianyulong* is found in a much more 'traditional' dinosaurian habit of forests.

## Stegosaurs

This is one of the rare groups of dinosaurs that needs no introduction: like *Diplodocus*, *Triceratops* and *Tyrannosaurus*, *Stegosaurus* is familiar to most people. These famous North American dinosaurs all had a sort of humped-back look with small forelimbs and larger hindlimbs, and they all had rather small heads, but when it came to their armour they were rather more complex. *Stegosaurus* rather famously had two rows of offset, large plates running down its back, then a cluster of four spikes at the end of the tail, but other stegosaurs had far smaller plates on the front half, and from the hips down the plates were replaced with symmetrical pairs of spikes. Not only that, but growing out from the shoulders was a huge pair of spikes that extended out from the body, making *Stegosaurus* in fact the least spiky of the stegosaurs (Fig. 14). While it is reasonable to infer that the tail spikes of stegosaurs would have provided some form of offensive structure (and research suggests that the tail could be swung in a wide arc, and powerfully, too),[12] quite what the spikes on the back would have done isn't clear. As armour, they really wouldn't have protected the head or flanks very well, and contemporaneous theropods could have bitten through them, while the idea that they may have been used as temperature control is controversial at best. The plates might best be thought of as some form of signalling structure. Stegosaurs had small and pointed heads and as a result were presumably rather selective feeders. Although they would have been able to rear up to reach foods located quite high up, they were presumably predominantly low grazers and browsers – though evidence from finds in the US suggest that *Stegosaurus* at least may have favoured drier environments.

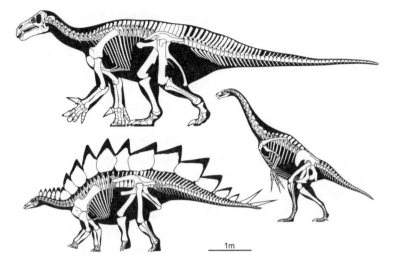

*Fig. 14 Assorted possible prey animals for tyrannosaurs although they would likely have focused on juveniles rather than the adults seen here. Clockwise from top: the iguanodontid* Iguanodon, *the therizinosaur* Nothronychus *and the plated stegosaur,* Stegosaurus.

Stegosaurs were at their peak in the Middle and Late Jurassic, and thus did not overlap much with most tyrannosaurs – and an adult stegosaur would have been a dangerous opponent for small tyrannosaurs like *Stokesosaurus*. Still, at least one stegosaur did survive into the Cretaceous in China, and might have met larger tyrannosauroids. In general, though, stegosaurs would probably not have featured on the tyrannosaurs' menu when still alive, since they were larger than contemporary tyrannosaurs and rather better equipped to defend themselves than most dinosaurian prey.

### Ankylosaurs

This group of armoured dinosaurs can be easily separated into two groups: those with a tail club and those without one. All ankylosaurs were pretty much coated in bony armour, with various plates, spikes and spines covering the body, typically in rows, to give them rather an upgrade in protection compared with their near relatives, the stegosaurs. Some of the ankylosaurs were almost encased in bony plates across the back, and in particular had arrays of spines that stuck out over the shoulders – though the belly was free and presumably vulnerable, but only if an enterprising predator could turn them over. Ankylosaurs also had skulls that were little more than solid boxes of bone, as these too were covered in pieces of bony armour fused to the bones of the skull to make an extremely tough head. Unlike the stegosaurs, ankylosaurs were rather low-slung animals that had extremely wide pelvises, making them broad and squat to look at. The tail clubs some bore would perhaps be better described as hammers since they actually had handles. The last few vertebrae were fused into a massive block of bone, often not unlike the 'club' symbol in a deck of cards, with three bulbous parts to it. Before that, however, was a string of vertebrae fused together into a solid rod, which is called the handle, and collectively this gave the tail great power and stiffness to produce something capable of delivering a powerful blow. It may have been used in fights with other ankylosaurs, but there's no reason to think that it could not have been deployed against threats from carnivores.

Unlike the stegosaurs, ankylosaurs were at their peak throughout the Cretaceous, and at the end of this period in particular many overlapped with large tyrannosaurs in Asia and North America, where their armour would have provided some serious protection from bone-crushing bites. Being low built, the ankylosaurs were also low browsers and grazers – though their heads were proportionally bigger and broader than those of the stegosaurs, allowing them to take in more food per bite and suggesting that they were less selective. Although ankylosaurs were typically rare components of faunas, *Pinacosaurus* from the Late Cretaceous of China and Mongolia was not only quite common, but also sometimes found in groups, suggesting that the animals may have often gathered together. Ankylosaur fossils are also often found in marine sediments, suggesting that they may have favoured coastal environments or estuaries, and even patrolled beaches looking for food.

## Basal hadrosauroids

These animals represent the early evolutionary radiation of the group that would go on to produce the hadrosaurs (better known as duck-billed dinosaurs), and includes a series of species that were one of the dominant herbivorous groups in the early Cretaceous in the northern hemisphere. Exemplified by *Iguanodon*, this group had many of the features that came to define the hadrosaurs. As one of the first dinosaurs named, and then the first dinosaur known from large amounts of material (coming from a coal mine in Belgium), *Iguanodon* has long been the subject of a great deal of research. The animals had beaks at the front of the jaws to help cut into vegetation, and, like the later hadrosaurs, they had rows of teeth in the mouth that would have effectively ground up their food, including tough plants like conifer needles.

They were capable of walking on all fours, with the tips of their fingers bearing hoof-like nails, but could also move onto two legs to rear up, and when moving quickly would run on just their back legs. This trait is termed 'facultative bipedality': they could shift to become a biped by choice, in contrast to, for example, the theropods, which were obligate bipeds, or sauropods, which were obligate quadrupeds. Like the later hadrosaurs, these animals were probably primarily quadrupedal (as many tracks show), though their rear legs were longer and stronger than their forelimbs. In addition to the specialised hooves of the fingers, many of them also had two rather interesting additions to the hands. First of all, the fifth finger (the 'pinkie') was offset at an angle to the three middle fingers, so that it could be folded across the palm of the hand; in essence it acted as something like a thumb, but at the bottom of the hand, not the top. This gave them some kind of limited grasping ability, which may have been used in foraging. The second, more famous feature, is the spike-like first digit. The thumb of *Iguanodon* and a number of its relatives was really little more than a sub-conical spike, which when topped with a claw would have made something of a weapon. This has been argued to have been used in defence against predators, or in fights between individuals over territory or mating rights. There's no real evidence for either hypothesis – though certainly an animal would have had to be almost in a bear hug with any theropod to be able to try and stab it, so the feature would have to be considered a last-ditch defence at best, and probably one of little value. The overall lack of a crest on the head – or other 'obvious' features – may have contributed to the fact that their taxonomy was for some time overly conservative. However, the group is now thought to be far more diverse than previously realised, and it is probable that like many other dinosaur communities, multiple closely related forms of it lived alongside one another in key locations.

## Hadrosaurs

In the northern continents of the Late Cretaceous, the duck-bills were the dominant clade of herbivore, essentially replacing the sauropods in this respect. Looking at modern faunas we see the artiodactyls dominate: cows, sheep, pigs, deer, antelope, and their relatives like camels and giraffes, are both diverse and numerous, and the hadrosaurs filled this role admirably at the peak of the time when tyrannosaurs lived on the planet. It is no surprise, then, that being the most numerous of the herbivores, they were also apparently the preferred meal of the tyrannosaurines. In general form they were close to their antecendents, capable of walking on two legs or four, and had long, beaked snouts and specialised rows of teeth in the jaws for slicing and processing plant matter. However, they did not have the gripping finger or spiked thumb, and the hands seemed to have been best suited to walking – though other uses cannot be ruled out. What did distinguish them, however, was the huge variety of crests on their heads. After the ceratopsians, they were the group of dinosaurs with the largest variety of exaggerated structures, which included

things that looked like giant helmets, horns and plates. Like
so many other unusual features, their exact function is not
known, but they seem to be have been primarily linked to
some form of sexual display or dominance signalling. In at
least one of the hadrosaurs, the giant *Parasaurolophus* from the
Americas, such a structure was connected to the nasal
passages; like some giant looped trombone, it could have
acted as a huge echo chamber to magnify the calls of the
animal.[13] Numerous modern species use sounds to
communicate, and the fact that many hadrosaurs are known
to have gathered together may have made this a valuable
means of signalling.

Mass nesting sites for hadrosaurs have been found, as well
as whole fields of footprints and mass mortality sites, showing
that they regularly gathered in groups, and the sizes of some
youngsters in the nests suggest that the parents took care of
their offspring for at least the first few weeks or months of
their lives. Although bony crests have dominated the
discussion when it comes to hadrosaur signalling, recently a
specimen of the apparently uncrested *Edmontosaurus* was
found with something like a cockscomb on top of the head;
a wonderful fossil replete with plenty of skin and scales, there
was also a fleshy extension at the back of the head. This
feature may have been widespread, and suggests that a number
of the apparently 'uncrested' hadrosaurs, while certainly
lacking any bony extension of the skull, might still in fact
have had some structure on the head.

There remains one common misconception about the
hadrosaurs, and that is that they were aquatic or semi-aquatic
animals. A popular idea among early researchers, it has now
been thoroughly disabused. The key idea was that their broad
tails would have been useful paddles in water, but in fact the
tails were bound together with numerous ossified tendons and
would have been rather inflexible: they could not have
functioned like a crocodilian tail as had been imagined. The
hands bearing hooves (and in a tight cluster of fingers, not a
broad and splayed paddle) also suggest a fully terrestrial
lifestyle, and the teeth were well suited to tough land plants,

an idea further supported by the finding of hadrosaurs with pine needles in their stomachs. Hadrosaurs could swim (almost all animals can), but the idea that they spent their days lounging in lakes, or could flee tyrannosaurs through the water where the carnivores would have struggled, are misplaced.

## Ceratopsians

Often known as the 'horned dinosaurs', many – though certainly not all – of these animals bore large horns on their faces. Most people will be familiar with animals like *Triceratops* and *Styracosaurus*, which were large-bodied quadrupeds with a massive head, bearing both forwards-facing horns on the nose or above the eyes and a massive frill of bone extending from the back of the skull and over the neck. Earlier ceratopsians, however, were quite small and may even have been bipeds, and they certainly lacked the horns and frills. These latter features were extremely variable, with all manner of different numbers, sizes and shapes of spikes, horns, bosses and other bits of bone. Key locations for them were on the nose, above the eyes, on the lower cheeks and around the edges of the frill. At least some of these animals fought each other using their weaponry,[14] and they certainly may have fought off, or at least threatened, potential predators and competitors – but their facial features may have been used mainly for social signalling and sexual display. Despite the diversity of these features, few were actually well suited for interlocking with or fighting other members of the same species, and the frills especially were too thin to have acted as a form of shield or armour as is often claimed, and were not nearly large enough to have served as effective radiators, as was once suggested. Several species of ceratopsian are known from mass mortality events, sometimes including dozens of animals, indicating that many may have spent time in herds, especially as young animals – though *Triceratops* at least may well have been solitary as an adult.

The earliest ceratopsians appeared in the Middle Jurassic in Asia, but in the Middle and Late Cretaceous they became one of the most common clades of the northern continents, and only the hadrosaurs were present in larger numbers. All ceratopsians were herbivores and had a combination of a sharp, pointed beak at the tip of the snout with rows of teeth behind, which collectively would have been able to handle many plants – though the overall build of the animals would have limited them to grazing or low browsing. There are a number of good skin impressions and pieces of fossil skin for the ceratopsians, showing that they mostly had small, pebble-like scales with occasionally larger ones in between. More intriguingly, a specimen of the small, Early Cretaceous *Psittacosaurus* had a series of long, filamentous plumes rising from the back and running along the tail. These filaments have been little studied, but they are similar in form to those seen in the heterodontosaur *Tianyulong*, suggesting that perhaps more ornithischians may also have had some kind of proto feather-like structures. Whether these were linked to feathers directly or had an independent evolutionary origin is still under debate. Other ceratopsians might have had something similar, and it has even been hinted that the larger scales along the backs of some later ceratopsians may have borne large, stiff filaments, giving them a slightly spiny appearance.

## Pachycephalosaurs

These animals were mostly small bipeds, yet the ceratopsians were their nearest relatives. Comparing an animal like *Pachycephalosaurus* to a *Triceratops* may sound ludicrous, but the earliest ceratopsians were at least occasionally bipedal, and lacked all the horns and huge frills of their descendants; one feature that unites these two clades is the development of a bony ridge at the back of the skull. In the case of the ceratopsians this became the huge and famous frill, but in the pachycepahlosaurs it developed into a massive hemisphere of bone on top of the head, often accompanied by some spikes and bosses around the edge, or projecting to the rear. The function of this dome has been argued about at great length, but it is largely agreed that it was used as some form of battering ram, though whether it was primarily used against other individuals of its own species or to fight off predators is unclear. Certainly, however, it would have packed a punch, and the larger animals were quite sizeable and probably relatively fleet of foot, so their primary mechanism for dealing with predators might well have been to flee, although they were hardly defenceless.

Collectively we know very little about this dinosaur clade: there are no complete skeletons, and what we find are mostly the solid bony domes. This suggests that the animals were rare and/or largely occupied environments with a low preservation potential: perhaps they only occasionally came to areas with bodies of water where they might have been preserved, but a number are known from tyrannosaur-heavy faunas. The mouths of pachycephalosaurs were small, implying selective feeding, though their eyes were large and they presumably had good vision.

## Sauropods

The sauropods are one of the great clades of the Dinosauria and are known from more than a hundred species, so in some ways it is difficult to generalise about them. However, on the whole these animals were large: even the smallest dwarf species weighed around a tonne, a great many weighed more than 10 tonnes, and at least a couple of dozen species were probably in the 30–50 tonne range, with some perhaps topping even that. Relatively familiar animals such as *Diplodocus*, *Apatosaurus*, *Brontosaurus* and *Brachiosaurus* provide a suitable blueprint for the sauropods as a whole. They were quadrupeds with deep chests, four long, columnar limbs, often a long tail, and in particular a long neck with a surprisingly small head bearing only a few teeth, mounted at the end. While the above figures give an indication of the size of these animals, some of the lengths involved are truly colossal (if often estimated owing to incomplete remains), in part because like the theropods, the sauropods were pneumatic, so there could be quite a lot of sauropod for not much mass: some of their bones were far more air than actual bone.[15] The Jurassic *Mamenchisaurus* from China had a neck reaching a staggering 13 metres in length, while some of the bigger sauropods may have lacked such a dramatically elongate neck, but could have had total lengths of around 30 metres or even more. The neck is the most notable feature of the sauropods and one that has attracted huge amounts of research.

Opinions as to quite what the neck may have been used for and how it may have functioned have varied over the years, but these days there is something of a consensus that its primary role was as an aid in food acquisition. A long neck would obviously help an animal reach up into the trees and take food that was high up, but it could also be useful for reaching forwards. Even taking a single step is quite a lot of effort for an animal weighing, say, 20 tonnes, but if it has a long neck the head can reach quite a swathe of area on one side or another. A great analogy is provided by geese that crop the grass while moving forwards slowly: watch them feed and you'll see that the head sweeps from side to side and covers quite an area as they walk.

One thing the necks of sauropods were not used for was to keep their heads above water: sauropods could swim, but they were not aquatic animals hanging around neck-deep in lakes and swamps (for a start, they would float rather than walk on the bottom). Their skulls were small and lightly built, and typically had only a few teeth that were often peg-like or suitable for nipping and stripping vegetation. Even so, their feeding strategy probably involved cramming down food in bulk, and relying on their large size and long digestive tract to get their energy. Indeed, sauropods (and some of the larger ornithischians) may well have benefited from their large size, which could have made them very efficient digesters, and they were perhaps able to get rather more energy from their food than smaller animals or comparatively large mammals like modern elephants. Their necks weren't the only odd feature, at least in some species, with some adaptations apparently having been 'stolen' from other lineages. Some of the Cretaceous forms were actually armoured rather like the early relatives of the ankylosaurs and stegosaurs, with small pieces of bone set in the skin, and even spikes around the shoulders. At least two different sauropods also independently evolved ankylosaur-like tail clubs with a large mass of bone fused together at the end of the tail – though without the stiffening handle.

The primary defence of these animals was probably simply being huge. They were typically slow and certainly not agile or normally armoured, but an animal weighing in the region of 10–20 tonnes would have been a danger to even very large predators. The sauropods appeared in the Triassic, but it was in the Jurassic that they truly peaked, with huge numbers of different sauropods occupying numerous ecosystems. In the famous Morrison Formation in the US there were perhaps 20 different species, all of which were multi-tonne animals. Their diversity took quite a tumble in the Cretaceous, especially in the northern continents, where they went from being the dominant clade of herbivore to one of the rarest. Even so, sauropods survived until the end of the Mesozoic, and the Cretaceous also saw what was probably the largest ever sauropod, the colossal *Argentinosaurus*. Because tyrannosaurs were largely small and rare in the Jurassic and sauropods were rare in the Cretaceous, these two lineages perhaps did not interact a great deal – though as we shall see, sauropods were on the tyrannosaur menu at least occasionally.

Moving away from the ornithischians and sauropods, there were several theropods that lived alongside the tyrannosaurs and would have offered tempting targets. Although the large tyrannosaurs would perhaps have been willing and able to kill and eat other carnivores, these are rather rare components of ecosystems and would not be a main prey item. Smaller dromaeosaurs, troodontids and oviraptorosaurs (and even the insect-eating alvarezsaurs) may certainly have been killed and consumed on occasion, but there were two major clades of theropod that were both large enough to be considered a decent-sized meal, and more importantly, being herbivores, were probably quite common.

## Ornithomimosaurs

The name ornithomimosaur means 'bird-mimic lizard', and while we do now know that a number of theropod clades were closer to the birds than the ornithomimosaurs, they were certainly very bird-like. More specifically, members of this group were very like ostriches, emus and other large flightless birds. They had very long legs, and a long, sinuous neck, on which was perched a relatively small head with large eyes and a beak – though they retained a typically dinosaurian tail. Flourishing in the north in the Cretaceous, fossil ornithomimids were found in large numbers in North America and eastern Asia, and they were quite diverse. Early forms had teeth, with *Shenzhousaurus* from China having numerous tiny teeth in the jaws that may have assisted it in feeding on plants or very small prey, like insects, while the bizarre *Pelecanimimus* from Spain had many long, thin teeth and a throat pouch, and presumably was some form of filter feeder. However, later forms seem to have been herbivorous, with a somewhat duck-like, rounded beak being combined with stomach stones to help in the digestion of plant food. The necks of these animals were long and flexible, and presumably helped them to reach the ground. They therefore probably fed and moved rather like ostriches, picking out small specific food items when foraging. It is not clear what function the arms had in this clade: they were rather long compared with those of earlier theropods, yet the claws were not overly curved or sharp, and the long neck would have served just as well, if not better, for reaching up or down to gather food.

A specific feature of these animals was their legs; ornithomimosaurs were among the fastest of the dinosaurs and would have been able to outrun almost anything else alive at the time. Like tyrannosaurs they had a stabilising arctometatarsalian foot, and the proportions of the legs overall, with a short femur, long tibia and long metatarsals, would have made them very quick. These features, combined with an overall small body and light build, would have made them fast runners. Estimates for top speeds vary considerably given the inevitable uncertainty, but it does seem reasonable to assume that they could achieve a speed of 60 kph, and probably more in bursts. Naturally, then, their most likely defence against predators was simply running away, though in addition their large eyes would have given them good vision, and they were also probably somewhat gregarious, with at least one large group of subadults being known, and this would also have helped them to avoid predation (the advantages of gathering in groups are discussed below).

Most ornithomimosaurs were of similar size, standing around 2 metres or more tall, but recent finds in Mongolia have led to final confirmation that the bizarre and little-known *Deinocheirus* was in fact a truly enormous ornithomimosaur with something of a humped and spinosaur-like back.[16] Until recently, this group had been assumed to have had feathers, with both earlier and later theropods being known with them, but feathered specimens were not actually described until 2013. Excitingly, these were from the extremely productive fossil beds of Alberta, Canada; it had been thought that the preservation of feathers would have been impossible here, but these finds suggest that many more such specimens might be preserved in other rocks. As in one of the oviraptorosaurs, the feathers found indicated that juveniles and adults had different plumages, just as birds do today, and that this trait was probably common in feathered theropods.

## Therizinosaurs

Almost certainly the most unusual of the theropod clades, therizinosaurs have both a wonderfully odd anatomy and an interesting history. The first fossil material found was from Mongolia, and was initially mistaken for some form of giant chelonian (the group of tortoises, turtles and terrapins), or at least an animal like that. Although there have been plenty of mistaken identities in the history of palaeontology, this stands out as a rather stark example; so what was going on? Well, the material was little more than part of an arm, and the most notable feature was the colossal length of the unguals: the biggest were more than 50 centimetres long. Large claws in a flat hand are something of a feature of chelonians, so this odd conclusion is not in fact quite as odd as it first sounds. The name eventually given to this animal paid tribute to this: *Therizinosaurus cheloniformis*, with the first half of the name having the evocative meaning of 'scythe reptile'. The rest of the animal, as was revealed by later finds, was also rather unusual, and it was most closely compared to the ancient giant sloths.

Therizinosaurs were bulky and heavy-set animals with rather short legs, and by dinosaurian standards rather short tails. The pelvis was especially robust and implies a heavy animal, though the neck was relatively long and with a small head. As in ornithomimosaurs, there was a beak at the front of the jaw, but all therizinosaurs retained teeth and had numerous small, leaf-shaped teeth in the jaws. It is the arms that were outstanding, however, though the hand of *Therizinosaurus* is apparently unique, and while there is quite a lot of variation in the manual unguals of therizinosaurs,[17] none of the others have quite such an array of giant claws. Oddly, though, these claws are very thin, and one cannot help but suspect that they were almost brittle, or at least vulnerable to breaking under heavy loads, and while they may have formed an effective threat display, they would probably not have posed a major threat to large carnivores. Thus the idea that they could fend off theropods seems wide of the mark; the commonly used analogy is with giant anteaters, which do rear up and brandish their claws when threatened, but these animals have much more powerful arms and very robust claws (used to dig into concrete-like termite mounds). Other suggestions for their use include raking leaves from trees, but quite how common these were in the Mongolian deserts – and the fact that the long neck would have already allowed the animals to reach into trees – makes their function something of a mystery. These animals form yet another feathered theropod group, although there is something rather special about the Chinese *Beipiaosaurus*. In addition to plenty of long filaments, this animal also had long, thick feathers along the back and especially along the neck, which were apparently stiff and rather more spine-like. These may have functioned as something of a defensive structure to put off would-be attackers.

In addition to the various dinosaurian species, the Mesozoic was replete with a huge variety of other animals. We tend to think of modern terrestrial environments as being dominated by mammals (and they clearly are to a degree), but that overlooks the birds, snakes, insects, crocodiles and other creatures that are there. Large tyrannosaurs would have favoured dinosaurian prey as they'd be after a proportionally large prey item, and that pretty much ruled out anything but another dinosaur most of the time – but even the biggest *Tyrannosaurus* was a baby once, and the smaller tyrannosaurs would not have been able to tackle many of the dinosaurs they lived alongside, young or old. A wide variety of animals we would consider normal components of ecosystems today were plentiful in the Mesozoic, and pretty much any meal available would have been taken at some point; meat is quite uniform and anything may be grist for the mill of the diet of a carnivore.

Just as today, there would have been lizards, snakes, tortoises and terrapins, frogs and salamanders, birds, fish, and various invertebrates like insects, millipedes and snails around that might have made the occasional meal for a tyrannosaur, especially the young of smaller species. For example, for the first year or so of their lives, modern crocodiles take little more than insects, tadpoles and fish fry, and it's quite some time before they graduate to big fish, let alone larger things like antelope. Mammals at this point in history were mostly very small, rat- or possum-like, and probably nocturnal, but that is not to belie their diversity. There were gliding forms and semi-aquatic swimmers, while *Repenomamus* was a predator the size and shape of a badger that was found in China with the remains of a baby dinosaur inside it,[18] and lived alongside *Dilong*. One can imagine that predator and prey may have swapped roles here, depending on who was the baby and who the adult.

Other dining options included the pterosaurs. Probably abundant in many ecosystems, these fliers would have been difficult targets, but we know of at least one that was eaten by a spinosaur, and a pterosaur bone has been described from inside a specimen of *Velociraptor*, so pterosaurs could have been caught by a lucky or enterprising carnivore. Continuing with the reptiles, there would have been a variety of crocodilians, not just the giant predators described earlier, but plenty of smaller forms and indeed many that were terrestrial and some that were probably even herbivorous. There were also reptiles such as the *Champsosaurus* (superficially very much like small crocodiles with very long snouts), and various other reptile groups that are no longer alive today but would have been potential targets.

## Stayin' alive

Key to the survival of any animal is avoiding being eaten by a predator. In the wild, few animals get to die of old age and any that suffer a serious illness or injury, or are beset with parasites, are likely to come a cropper as a result of being too slow, or too slow-witted, to avoid being killed. However, as mentioned above, herbivory is a tough life to lead, and herbivores typically either have very broad diets, consuming anything and everything available to them (cows are rather indiscriminate grazers, for example), or are selective and pick out only the best foods, substituting quantity for quality (giraffes are rather delicate feeders in this regard). Either way, though, this takes up a lot of time, and can require long hours of feeding or searching respectively; while a herbivore is looking down and feeding on the ground or deep in a bush, or high up a tree, it's difficult for it to look out for something aiming to make a meal of it. Keeping an eye out is therefore key, which is why feeding animals of all kinds regularly stop what they are doing and put their heads up to look around for any potential threats. Although getting a meal is very important for a prey animal, and it can't afford to miss endless opportunities, one mistake may mean the end of it.

Vision is often the key to spotting threats, and dinosaurs did have good vision in general; many species, like the ornithomimosaurs, had large eyes. However, smell and hearing would also allow for detection of threats, and they can be more omnidirectional than visually spotting for predators. Our understanding of these senses is more difficult to calculate: orbit size is a good correlate of eye size and by extension visual acuity, and we can get an indication of the importance of smell from the shape of the brain, and the range of hearing from the structure of the ear, but the details may be rather sketchy. An animal may have a poor sense of smell overall, but if that is finely attuned to the scent of a tyrannosaur it might be most useful.

One key aspect of vigilance and defence is collaboration, both with other species, and especially with members of the same species. Plants are typically fairly plentiful, so a herbivore may not have that much competition with its immediate neighbours overall, and what you have can be more than offset by the ability to defend yourself against attack. Obviously, animals of the same species will have the same habits and want to do similar things at similar times, so are already likely to cluster together. Though this doesn't necessarily mean that they are social, being part of a group does offer a number of benefits. First of all there's a dilution effect: if a carnivore chances across a prey animal when it is alone, then this is the sole target of an attack. In a group, however, there may be dozens of other animals, and there's a good chance that the attacker will target another animal, perhaps an individual that is ill or otherwise vulnerable. Sheer numbers may have an effect in reducing the effect of predation, and certainly massed ranks of defenders can present a united front. One sheep alone may not be able to do much when faced with a wolf, but a group all prepared and able to charge, and from multiple angles, makes a far more imposing and potentially dangerous proposition.

More importantly, though, where there are multiple animals together, the time needed to spend looking for predators is hugely reduced. Less time needs to be spent in looking around, and there is a much greater chance of one of the group spotting a threat. Logically, this makes sense, and studies in the field show that where predators are more common, more solitary animals tend to form groups, while the sizes of existing groups increase. Gathering together is thus an important aspect of prey behaviour, and it's not surprising that many prey species of the tyrannnosaurs probably spent large amounts of time together, especially where food was plentiful.

Being part of a group is particularly important for juvenile animals, which suffer a huge swathe of penalties compared with adults when it comes to finding food. First of all, they are growing and thus need proportionally more food than adults, which entails longer foraging times. On top of this, they are typically poor foragers: being young and inexperienced they are often poor at finding food, and adults, being larger, are able to keep them off the best places to feed. This limits the juveniles to poorer quality food, which they then have to eat more of to meet their nutritional requirements. All of this entails longer times feeding or longer searching times, and perhaps also searching in areas where danger is more prevalent; juveniles are therefore much more vulnerable to predation than adults. Add to this their own naivety about predators (they may not recognise predators they have not encountered before, or act on signals from other animals), and they become exceptionally vulnerable. It is perhaps not surprising, then, that there is very strong evidence not only for groups of juvenile dinosaurs having aggregated together, but also that this may have occurred in some species where the adults were primarily solitary.

Detecting a predator is just the first step in surviving, predators also need to be avoided. Naturally, foraging in sensible areas (keeping away from classic ambush sites such as dense cover in trees, or staying downwind of such cover so trouble can literally be smelt coming) goes a long way to avoiding being targeted in the first place, but there will be plenty of occasions when an animal is actively attacked. The vast majority of animals would flee as a first response, among living species even those capable or well equipped to repel predators usually run first. Certainly, many dinosaurs were well equipped in this regard, even allowing for the speed of many tyrannosaurs, and ornithomimosaurs, hadrosaurs and heterodontosaurs would doubtless have run, while pterosaurs would have taken off to avoid predators. At least one *Triceratops* specimen seems to have had a face-to-face encounter with *Tyrannosaurus*, with evidence of tooth marks lying along the horn of the ceratopsian.[19] However, this may not have been the normal defensive strategy for ceratopsians, for a start, many were too small to utilise it, and didn't have the large and forwards-facing horns that *Triceratops* possessed. Additionally, a predator would not generally have wanted to face such a well-armed opponent, so such examples are likely to indicate exceptions rather than the rule. No carnivore would last long if its main approach to getting dinner was to charge an animal so well equipped, and hope to survive the kind of dramatic battle that forms the mainstay of dinosaurs in cinema; the old, the sick and the young are the classic prey of choice. How to actually get them, though, is another matter.

# CHAPTER FOURTEEN

# Competitors

The various tyrannosaurs were never the sole carnivores in their respective environments, and for most of their history on Earth they were not the largest carnivores either. At the end of the Cretaceous in North America and eastern Asia, the tyrants were the dominant carnivores (though there was typically more than just one tyrant species in any given ecosystem) – but there was plenty of competition from other theropods and indeed other reptiles that were around in the Mesozoic. Evidence of any direct interactions between the tyrants and these other animals (such as for combat between them, or for multiple species having fed on the same carcass) is lacking, but these would undoubtedly have occurred at some point. There would also have been indirect pressures that would have affected both the tyrants and the competing species; carnivores do not exist in evolutionary and ecological isolation, and every piece of land staked out, and even each animal preyed upon by one species, is going to affect others.

The rarity of carnivores in terrestrial ecosystems is ultimately determined by basic energetics. Plants, even at their best, can only trap around 2 per cent of the incoming energy from the sun. Of the energy that is captured, some can be used by the plants for growth and to produce the next generation, and much of it is used to keep them alive using basic biological processes. At best, around 10 per cent of the energy can make its way into any herbivores feeding on them, and again only 10 per cent of that will be transferred to any carnivores feeding on the herbivores. A tyrannosaur would only have really been in a position to tackle relatively large prey with any regularity (fun though it is to consider, no adult *T. rex* would have been able to survive on a diet of beetles or mice). Plenty of smaller herbivores like insects also take energy from the available plant matter, so it's easy to see why there wouldn't be a multi-tonne tyrannosaur every kilometre or two in the Cretaceous.

If dinosaurs, especially larger forms, were either exploiting their large size to keep warm (rather than directly burning calories like mammals do), or were simply ectotherms, more carnivorous individuals could have existed. If animals are not using a great deal of their consumed energy to keep themselves going, then for the same amount of plant matter there can be more herbivores and therefore also more carnivores. It's pretty much impossible to calculate such ratios from what is retained in the fossil record, but regardless, the carnivores would have been a proportionally rare component of the fauna whichever way you cut it.

Even though carnivores may normally be relatively uncommon, there are generally multiple large forms of carnivore in any environment. Looking around today, we typically see a combination of dogs, cats, bears and/or hyenas in almost every major terrestrial ecosystem (Australia and Antarctica aside, for rather different reasons). Although they come and go in various places, you can easily see lions, one or two hyenas, cheetahs, leopards and hunting dogs in much of Africa, leopards, tigers and bears in most of Asia (and lions and hyenas in parts of India, as well as wolves and dholes), and bears and pumas across a great deal of the Americas, with wolves in the north and jaguars in the south. Although they are somewhat varied in size, types of prey taken and behaviour, they are nonetheless large carnivores, and the step down from them is often considerable (things like various jackals, bobcats and badgers are often barely half the size of these large forms).

People often talk about 'top' or 'apex' predators in a given ecosystem, but the term is often misappropriated when it comes to animals like lions, or for that matter tyrannosaurs. In ecology, 'top predators' refers to what are generally the largest of the carnivores in an ecosystem (or, more formally, those that occupy the highest trophic level), but more specifically this term is used to distinguish such animals from *other* carnivores that they themselves feed on. For example, large sharks are top predators because they eat largely other predators (large fish and sea lions, which are themselves also predators, eating smaller fish, squids, crabs and the like). So in this context, the term 'top predator' for a lion is incorrectly applied. Although lions do kill and eat other predators, the vast majority of their food comes from herbivores, and they are no more a top predator in their ecosystem than are jackals or even mongooses: they all occupy the same ecological level. Still, these animals may or may not be in competition with one another, relationships between animals in food webs being rather complex.

Competition between animals may be direct or indirect. In a direct interaction two different species of animal (individuals or groups) are in conflict over a specific prey item or resource. For example, a *Gorgosaurus* and an *Albertosaurus* may battle over a dead hadrosaur, or both might try to drink from the same waterhole or use the same nest site. Only one of them is going to win, and the other will suffer directly as a result. The animals could also compete indirectly, that is without ever actually coming into direct contact with one another. One could hunt only at night and the other in the day, but every animal eaten by a *Gorgosaurus* would be one out of the pool of available prey for *Albertosaurus*.

Such ecological interactions are common, and can cause major issues. The introduction of a new predator to an ecosystem can cause havoc for the stability of the system, with some species benefiting enormously and others becoming extinct as a result. For example, due to the reintroduction of wolves to Yellowstone National Park in the US, coyote numbers have fallen, but both bears and small carnivores such as foxes have increased. The number of red deer has dropped dramatically (though the number of available red deer carcasses has gone up), and their habits have also changed as a result of their encounters with the reintroduced carnivore. However, the reintroduction has benefited other herbivores, with bison, beavers and small birds all doing well and so providing benefits for the other carnivores. The effects of such changes to an ecosystem can be very far reaching and sometimes difficult to predict, and it is necessary to know which species are around, and how they might be interacting, to understand them. It's obviously very hard to tell how the dinosaurs interacted, we're generally limited to snapshots in time, and rarely find evidence for interactions between the carnivores and their prey, let alone with each other. However, with typically only a handful of large predators in any given ecosystem, they must have been competing at some level.

### An ever-changing and evolving roster

Just as the various lineages of tyrannosaurs were moving between continents and changing in shape throughout the Mesozoic, so too were their competitors. Thus while, say, *Stokesosaurus* and *Dilong* had quite a few non-tyrannosaurs to compete with (or indeed worry about) in their respective habitats, *Tyrannosaurus* apparently stood all but alone. There was naturally something of a transition: before the larger tyrannosaurids evolved, no tyrants were among the largest carnivores in their respective ecosystems, but in the latter part of the Late Cretaceous they were always among the biggest theropods wherever they were present.

The competing lineages of carnivorous dinosaur varied considerably in their anatomical refinements, with each lineage being specialised according to the way it targeted its prey. The spinosaurs, for example, had crocodilian-like skulls and are known to have eaten fish as well as other dinosaurs, the carcharodontosaurs had much narrower heads, with thinner, sharper teeth than tyrannosaurs, and the azhdarchids were toothless and considerably different from the rest of these groups in that they could fly, being large pterosaurs, not theropods.

### Ceratosaurs, megalosaurs and allosaurs

Although each of these groups represents a separate radiation of carnivorous theropods, it is convenient to treat them together because they have generally similar body plans, and by extension probably operated as carnivores in similar ways. The megalosaurs were generally the smaller bodied of these clades, and the allosaurs were both longer lasting and became more diverse, as well as including some real giants in their ranks, while the ceratosaurs were an early theropod clade that went on to produce the bizarre abelisaurs. All had relatively long and powerful arms, and proportionally narrow skulls with thin, sharp teeth, so they were in many ways archetypal theropods, exemplified by animals like *Megalosaurus*, *Allosaurus* and *Ceratosaurus*.

Given the size of their arms, these animals presumably used them when acquiring prey, and while the teeth would have delivered a strong bite, the skulls were much weaker than those of the tyrannosaurids, and the teeth were thinner and more numerous, suggesting more of a cutting bite than a crushing one. *Allosaurus*, for example, may have used the head as something like a saw-toothed hatchet, slashing at prey and removing chunks of flesh, or leaving large, bleeding wounds.[20] The *Megalosaurus, Ceratosaurus* and earliest allosaurs would have overlapped with some of the early tyrannosauroids, while later allosaurs like *Allosaurus* and the large and crested *Ceratosaurus* overlapped with animals such as *Stokesosaurus* (in North America, at least). A later group of allosaurs, the carcharodontosaurs, got especially large and contained some of the largest terrestrial predators of all time. Like *Spinosaurus*, several of these animals exceeded even *Tyrannosaurus* in length, but probably not in mass. The most famous and indeed largest members of the carcharodontosaurs herald from the Late Cretaceous of South America, and thus failed to overlap with the tyrannosaurines and albertosaurines (sadly for enthusiasts of 'who would win in a fight between' scenarios). In the Early Cretaceous, however, they did overlap with the tyrannosaurs in Asia and the US, and while we have yet to find a locality that definitively contained both, there are places that are close together in time and yield fossils of one or the other.

Interestingly, there does seem to have been something of a changeover of the carcharodontosaurs and tyrannosaurs, with the former clade providing the larger and more numerous fossils in the Early Cretaceous and first part of the Late Cretaceous, before the more derived tyrannosaurs took over and the carcharodontosaurs fell away. It is not clear what may have prompted this turnaround, but perhaps environmental changes favoured the tyrannosaurs and the carcharodontosaurs couldn't adapt or keep up with the changes, or maybe the tyrannosaurs were superior in some way, but it took many millions of years for this to happen. My personal suspicion is that the faunal turnover of the time was responsible, because at this point the ceratopsians took off, and the ankylosaurs and hadrosaurs were also diversifying. Changes in prey types would probably have had a serious impact on the major carnivores eating them, and it is easy to imagine that some aspect of the behaviour or biology of prey animal groups favoured tyrannosaurs over carcharodontosaurs. Bear in mind that at this point, the giant heads and teeth of derived tyrannosaurines and albertosaurines had yet to appear, so these two clades were rather more similar to each other than the last forms of each, suggesting a greater degree of subtlety in their approaches to getting a meal.

## Spinosaurs

Members of this clade are perhaps the most interesting and unusual of all theropods, as they look quite different in a number of ways from any other clade. They were rare, and although the best fossil material has come out of North Africa and Brazil, a superb skeleton of *Baryonyx* (Fig. 15) has been found in the UK, and there is evidence of the animals in East Africa, Thailand and central China. Almost all the fossils are from the Cretaceous, but given their evolutionary history spinosaurids must have been around in the Late or even Middle Jurassic – although the fossil evidence for this is lacking, perhaps because they were so rare in most environments.

The name of this group comes from *Spinosaurus*, the charismatic and truly giant theropod known from Egypt and Morocco, which as its name suggests had a huge 'sail' of elongate dorsal spines along the back that were up to 2 metres tall. Other spinosaurs were not quite so well endowed, though

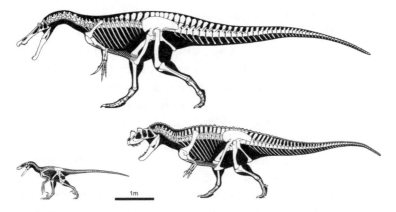

*Fig. 15 Several carnivores that tyrannosaurs may have competed with (or been threatened by). Clockwise from top: the spinosaurid* Baryonyx, *the ceratosaur* Ceratosaurus, *the dromaeosaur* Deinonychus.

they would have had something like a large ridge along the spine. The lack of complete material for this group has made estimating the animals' size difficult, and indeed contentious, with some estimates for *Spinosaurus* reaching a colossal 17 metres,[21] and while this is likely to be on the high side, equally it is perfectly plausible. This would make this genus potentially rather larger than *Tyrannosaurus*, at least in terms of total length, though mass estimates would suggest a lighter animal overall; the big tyrannosaurs were really bulky and solidly built. Giant sails aside, spinosaurs also stand out as having had rather strong arms and in particular huge claws on the hands. They did not have the rather reduced arms often seen on large theropods, though quite how they may have used their arms is not clear.

The real standout piece of anatomy for a spinosaur was the skull. To look at, the skulls of spinosaurs appear rather like those of modern crocodiles: they are long and low, with a rather uneven look to the edges that hold the teeth, the tip of the jaw is apparently expanded, and there is a distinct notch where the premaxilla meets the maxilla. Looking inside there is a very unusual feature: a secondary bony palate, so that essentially the roof of the mouth is duplicated (another feature also seen in crocs), and their teeth are near conical and much more like those of the crocodilians than the teeth of other theropods. All in all, it's quite a suite of features, and these collectively point to a crocodilian-like method of catching food. Teeth with a circular cross-section would have been strong and resistant to breaking, and also good at puncturing prey and holding onto it, and the secondary palate provided excellent resistance to torsion of the jaw. In short, the animals appear to have been well equipped to grip struggling food items, and if necessary shake them to kill them. These rather crocodilian adaptations also suggest that fish might well have been on their menus, and indeed *Baryonyx* is known to have eaten fish, suggesting that spinosaurs may have been more suited to this dietary option than other theropods. However, crocs also tackle terrestrial prey, and the specimen of *Baryonyx* mentioned above also shows evidence of having eaten a young iguanodontian. Spinosaurs may have occupied habitats where few other theropods could thrive, and because they fed on a wide range of prey types they could have survived where others would have struggled to find sufficient food. If this is the case, they may not have encountered the tyrannosaurs much, even where they did overlap.

The different theropod lineages were not the only ones to have supplied large-bodied competitors to the various tyrannosaurs of the Jurassic and Cretaceous – other dinosaurs were also around to get in on the act. Just as today large carnivorous birds such as eagles and vultures may take food (almost literally) from the mouths of big cats and wolves, so too in the Mesozoic large pterosaurs were in the air and looking for food. Similarly, the waterways of the dinosaur era had their share of aquatic and semi-aquatic reptiles. Today, crocodilians are rather limited in number and diversity of form, but formerly there were many more kinds of crocodile, including those that lived almost exclusively on land, and others that were marine and rarely left the water. However, there were some that were similar in overall form and habits to the larger crocs we share the world with today. They were big-bodied animals that could swim well but also ventured far inland, ranging from coastal waters to rivers and lakes, and they had the capacity and tendency to tackle large prey.

## Azhdarchids

The azhdarchids were the largest of the pterosaurs, reaching a peak in the Late Cretaceous, where they overlapped with the largest of the tyrannosaurs across the northern hemisphere. Some had a staggering 10-metre plus wingspan, more than triple that of the wandering albatross (the bird with the biggest wingspan today) and closer to a small plane than anything biological flying in the air these days. Pterosaurs are often incorrectly considered to have been flying dinosaurs or ancestors of birds, whereas in fact they formed their own archosaur lineage – though one close to the dinosaurs. Common perception also sees them as poor fliers, capable only of gliding or soaring, and ungainly on the ground and more like giant, inept bats than anything else. However, this is the polar opposite to the truth – they were actually superb aeronauts that were not only the equals of birds, but perhaps even their superiors when on the (membraneous) wing. Similarly, they were every bit as competent as birds on the ground or, more accurately, as nimble as the incredibly agile and active vampire bats when grounded. They were quadrupedal, folding their wings alongside the body, and walking and even running around on all fours like rather odd reptilian giraffes. In this configuration, unlike birds, they could use their huge flight muscles to take off (birds actually jump using their legs, then flap with the wings), so could launch into the air in just a second, and did not need the long run-up that birds such as condors and swans require.[22]

Despite their huge size, they were not actually that heavy; due to their pneumatic skeletons and other adaptations, even the biggest azhdarchids weighed only around 250 kg, which is truly incredible for an animal that would have been intimidating even for a large tyrannosaurine; after all, they had a head (which was mostly beak) approaching 3 metres long. Direct evidence for the diet of azhdarchids is lacking, but their sharp beaks indicate that they fed on meat. They were perhaps the best adapted for walking on land of any pterosaur clade, and their remains are found primarily in terrestrial environments, not out to sea like those of most other pterosaurs. The best analogy for them may thus be birds like maribou storks and ground hornbills: large birds that spend most of their time on the ground, but which hunt down small mammals and reptiles, or scavenge from dead animals.

## Crocodilians

Several huge crocodilians lived alongside the tyrannosaurs at various times, though the most obvious competitor for them would have been the giant *Deinosuchus*. This monster could reach sizes considerably greater than those of even the largest of modern crocs, with estimates running to up to 10 metres long. As for many dinosaurs, there are few good skeletons of this animal, and most fossil material comes from teeth and bony pieces of armour from along the back.

Despite the variety of crocodilians in the Mesozoic, *Deinosuchus* and its nearest relatives were in fact very similar to modern crocodilians in terms of anatomy and, by extension, behaviour. *Deinosuchus* is actually from the evolutionary branch that includes modern alligators, so it was closer to these than the other living crocs, including true crocodiles. In many ways *Deinosuchus* were really giant alligators, being semi-aquatic, and they would have been able to eat pretty much anything in the water or on land. While the primary diet may well have been fish and terrapins or turtles, they would have been more than capable of taking down dinosaurs. Indeed, shells of turtles with damage attributed to *Deinosuchsus* teeth, and even some hadrosaur bones with bites on them, suggest that such events were not uncommon. When attacking large animals on land, these crocodilians would have been ambush predators: striking from the edge of the water and trying to drown their prey. Occurring only in the latter part of the Late Cretaceous, *Deinosuchus* has been found from northern Mexico to Alabama, in the US, so would have overlapped with a number of later tyrannosaurs. Other giant crocs are known at this time from North Africa and other locations, but these were places apparently devoid of tyrannosaurs.

So far this chapter has focused on the larger theropods and other archosaurian carnivores that would have been direct competitors of the various tyrannosaurs as adults. However, juvenile tyrants and smaller animals like the proceratosaurids would also have faced competition from some of the smaller theropods, and from smaller pterosaurs and crocodilians. Even if, say, large tyrannosaurines did exhibit extended parental care for several years, that could still have resulted in relatively small and independent young tyrants having to forage for themselves and coming up against other small theropods, directly or indirectly.

In particular, the various bird-like theropods that are close relatives of the avians were a collection of small carnivores that would have taken a variety of foods, perhaps including small or young tyrannosaurs. These animals were all completely coated with feathers, and if around today they could easily be mistaken for large birds at a distance. At this point in the theropod family tree, the difference between non-avian and avian dinosaurs is pretty blurred, and some species have even flitted between the two sides of this line in various cladistic studies. A great many of the features we associate with living birds can be seen in these groups, especially the dromaeosaurs and troodontids, which are particularly close to the birds. Like the tyrannosaurs, some of these animals have been the subject of considerable speculation. Enthusiastic repetition of unsupported ideas by the media, and their appearances in pop culture (*Jurassic Park* has a lot to answer for when it comes to *Velociraptor*), has given them a rather unbalanced reputation.

## Oviraptorosaurs

These are perhaps the most interesting of the small theropods and form a group that certainly included a number of herbivores in its ranks, and probably some omnivores or carnivores. They are the theropod clade with perhaps the most interesting name going, and one that has a rather nice twist to its origin and use. The name of the clade derives from the first-discovered animal: *Oviraptor* translates as 'egg robber', and the name was coined based on the common association of the remains of these dinosaurs with nests of eggs in the deserts of Mongolia. However, we now know that these animals were not actually stealing the eggs of other dinosaurs, but were parents, brooding their own eggs.[23] They were clearly devoted parents, which died while protecting their eggs, perhaps because they were overwhelmed by a sandstorm or similar event. That said, I'd be most surprised if some of these animals didn't take eggs when the opportunities arose, leaving them with an actually appropriate name.

Oviraptorosaurs were mostly toothless, and instead of teeth had deep and rather parrot-like beaks that might have served equally well in shearing tough plants and seeds, and cutting through small dinosaurs and other prey. One nest contained the remains of a lizard that may have been the remains of dinner for an oviraptorosaur, or was perhaps brought to the nest site for the hatchlings to feed upon. One giant Mongolian oviraptorosaur from the Late Cretaceous would have rivalled *Alioramus* in size. Its diet is unknown, but even if it was a herbivore, it might have been one for the tyrants to avoid.

## Troodontids

This is another group of animals with contentious dietary habits, as some researchers have suggested that they may have been primarily herbivorous. This is in part because the serrations on their teeth are absolutely huge and unlike those normally seen on theropods, and are more similar to the giant serrations often seen in the ornithischians, which used them to cut up leaves. However, leaving the teeth aside, troodontids clearly had a number of carnivorous traits, including long claws on the hands and feet. Like in the dromaeosaurs described below, the second toe of each foot of a troodontid was short and was normally held raised off the ground (we see this in the posture of the foot in fossils, but it also shows up in two-toed footprints). Just as in cats, this adaptation kept the claws from being blunted on the ground during day-to-day activities, and kept them sharp for when they were needed. In short, the troodontids possessed a rather lethal weapon.

These animals were all rather small, with few reaching even 10 kg in mass, yet they might still have been dangerous to a small tyrannosaur. At least one had offset ears, rather like those of modern owls, which would have given them superb directional hearing, and combined with their rather large eyes this suggests that they may have been nocturnal, or at least able to hunt in low light, though presumably they tended to target small prey. Their namesake, *Troodon*, is well known for having the proportionally largest brain of any dinosaur; they may not have been especially brainy, but clearly they were not likely to have been dumb, either.

## Dromaeosaurs

The dromaeosaurs were certainly all carnivores, and with a wide variety of potential prey. Various fossil specimens have been found with bones in their stomachs from fish, birds, mammals and pterosaurs, and evidence suggests that they would have both scavenged and actively hunted their meals. They have been described as perfect killing machines, and while this assertion clearly involves at least a bit of hyperbole, it's easy to see how this reputation came about. Like the troodontids, the dromaeosaurids were typically fairly small for theropods (few were much bigger than 2 metres or so, and many were considerably smaller), lightly built, and with long legs and proportions that suggest that they were fast runners. Unlike their cousins, they had a specially stiffened tail that presumably helped give them greater balance. In addition to the usual array of teeth, they had long arms with large claws on them, and like the troodontids, a large and retractable claw on each foot. Unlike in the troodontids, however, this claw was especially large and suggests a rather specialist use of this feature. Quite what that may have been has been contentious, with a number of suggestions being put forwards, all ultimately linked to predation.

A key issue with regard to dromaeosaurs is whether or not they hunted in groups, and if so, whether they targeted large prey. This exciting hypothesis was mainly sparked by the discovery of multiple remains of the famous dromaeosaurid *Deinonychus* with bones of the ornithischian *Tenontosaurus* – but this is mostly the limit of the evidence in support of the hypothesis, and it's not especially convincing.[24] You wouldn't expect a group of predators to die alongside their meal, it's not a good long-term strategy to try and hunt something that can kill multiple members of your group, so this find appears to represent an oddity rather than a caught-in-the-act moment. Just finding some animals together once does not mean that they were social pack hunters, and a dead herbivore was not necessarily killed by them. This is not to rule out social hunting, but it is no more convincing a case than there is for social tyrannosaurs; the idea that these animals were specifically going after large prey as a group really has no strong evidence in its favour.

### Room for one more?

At various times in the tyrannosaurs' evolutionary history, this roster of Mesozoic carnivores would have posed greater or lesser issues for them. The tyrants of the Jurassic were relatively small and perhaps only faced direct competition from smaller or juvenile megalosaurs and early allosaurs; moving into the Cretaceous, the carcharodontosaurs became an issue, and the rise of the great crocodilians and azhdarchids brought new competitors to the fore; at the end of the Cretaceous, the biggest problem was typically other tyrannosaurs. In central China, *Yutyrannus* and *Dilong* shared an ecosystem; in eastern China, *Zhuchengtyrannus* lived alongside at least one other tyrannosaur; in Mongolia, *Tarbosaurus* would have seen *Alioramus* and *Raptorex*, while in the Americas, first *Lythronax* probably overlapped with *Bistahieversor*, and later *Daspletosaurus*, *Albertosaurus* and *Gorgosaurus* were probably in contact with each other at various times. In short, there was variety, and even *Tyrannosaurus*, often considered a lone outsider in terms of competition from other theropods, may have had the odd meal taken by a giant azhdarchid, or rued the loss of some baby hadrosaurs to a troodontid or dromaeosaur just 1 per cent of its size.

In ecological terms, however, the various carnivores must have been separated to some degree. True, there was competition, but if two species have exactly the same requirements, the assumption is that one will eventually drive the other to extinction (by dying off or being forced out of the environment). One species may be a little better at taking the food than the other, reproduce faster or have some other slight advantage. Of course, the ever-changing world (including such factors as bountiful summers, occasional droughts, new species arising and the spread of disease) would keep everything unstable enough for multiple direct competitors to survive, but probably not if they had exactly the same habits. What we would therefore expect is some degree of ecological separation: two competitors taking different routes to getting enough food, so that neither is too restricted by the other.

There is great variety in how animals do this. Any two competitors might target different prey species, or different members of a given prey species (such as juveniles versus adults). They may hunt in different places (by the water or in the forests), hunt at different times of the day, or even at different times of the year (if they or the potential prey animals migrate), or they may be predominantly or exclusively scavengers rather than active predators. Any of these factors might be enough to keep two species separate enough to avoid one excluding the other through competition, and more than likely some combination of them was operating at any given time among the dinosaurs. Moreover, such traits would probably vary in different parts of the ranges of these various animals; where a species overlapped with several competitors it would have to focus more on some foods than it did in places where there were fewer or no competitors. Individual variations can be quite impressive: one individual might only hunt a certain species and at night, while another might go after a different species in the day. Evolution hones features and behaviour towards certain specialisations (tyrannosaurs have a powerful bite, while spinosaurs can target fish effectively), but these can be extremely variable, and it's why there are such oddities in the world as antelope that hunt and eat small birds and mammals.

The major differences in anatomy between these groups, and the evidence available from stomach contents, tracks and bite marks, point only towards general suggestions as to how they may have separated themselves ecologically from each another. Inevitably, there are things we cannot easily tell apart, for example, despite some claims that various dinosaurs were nocturnal or diurnal based on the size and shape of the orbit, this is not an especially reliable indicator of typical habits, so things may have been rather more complicated than we imagine. Habitat differences are also very hard to detect, and animals that lived well away from water would probably rarely have been buried, so would be rare in the fossil record. We might think of them as being rare animals as we have so little material, but they might have been very common, just living somewhere else. Critical, too, is the fact that different animals hunt in different ways. In Africa, lions, hunting dogs, hyenas and crocodiles all regularly catch various antelopes and do it in different ways, but for this resource at least they are in competition, and just because one species is ambushing animals from the water and others are charging after them across the plains, does not stop them interfering with each other. In short, biological interactions are complex and subtle.

Still, based on the available data we can draw some broad conclusions about the different habits of the tyrannosaurs' competitors. The crocodilians (or at least, the large carnivorous ones) would have been rather like those of today: unable to cover long distances on land quickly, they would have relied on ambushing prey from the edge of the water. Animals that came to drink or were forced to cross bodies of water would have been particularly vulnerable, and one large bite might have been enough to kill potential prey, or they might have been dragged into the water and under the surface. Similarly, the spinosaurs were clearly active close to or in the water, but would have been much more capable on land than the average big croc. It has been suggested that they tended to occupy environments with limited resources, where their ability to forage effectively both on land and in water might have enabled them to get enough to eat, while animals limited to one or the other might have struggled. In contrast, the azhdarchids would have stuck to open areas; although when walking they had surprisingly compact frames, it's probable that animals with 10-metre wingspans avoided forests and cluttered environments as it would have been hard for them to take off from them, or to manoeuvre through them. So these were most likely to have been creatures of the plains, snapping up small prey such as young dinosaurs or lizards, though this doesn't mean that they didn't take aquatic prey or forage along shorelines.

The plethora of dromaeosaurs and troodontids in fossil beds known to represent forested environments might be a result of a preservational bias for small animals in these beds, or point to the fact that these animals did well in such places. Both, of course, may be true, and certainly it is easy to imagine these small and nimble animals to be well suited to ducking between branches and scrub in a way the giant theropods would presumably struggle to match. Still, these smaller carnivores are also found in abundance in deserts, so clearly they were not limited to forested environments. Even if we allowed for the possibility of pack hunting, most of these animals were rather small, and it is difficult to accept that even a large group would regularly bring down very large dinosaurs, so the majority of their prey was probably small and also varied.

In the case of the larger and terrestrial theropods, clearly some could operate in cluttered spaces. *Yutyrannus* was not a small animal and the fossil evidence suggests that it lived in an environment that encompassed a great deal of forest. Presumably similarly sized theropods from other clades may have operated in closed spaces, though of course large prey might also have shunned these areas, which could have been a complicating factor. The allosaurs, carcharodontosaurs and megalosaurs occupied a number of environments, as exhibited by their wide range and varied ecosystems, so (as is the case for many modern large predators) it is hard to characterise any group as a whole as being specialised for plains, forests or other habitats.

At least some allosaurs may have attacked prey with heavy, slashing blows from the head, the intention presumably being to try and bleed out their prey rather than dispatch it with one massive blow (though this may have been effective against smaller animals), and the later carcharodontosaurs may have followed suit. Certainly, they had relatively sharp teeth and were not as suited to heavy bites as were the tyrannosaurs (not that this made them incapable of a strong bite), and their narrow heads would not have allowed them to take large bites, either. Quite what the megalosaurs were doing is really not clear, and little work has been done on their habits. None of the animals in these groups appears to have been especially quick: the proportions of their legs are not like those of the superfast or efficient runners such as alvarezsaurs and ornithomimids, and for that matter, many tyrannosaurs. They may not have been well suited to quick sprints after prey and relied on something closer to an ambush to tackle it.

This, then, covers the competitors of the tyrannosaurs and their probable interactions. But how did the tyrants themselves operate? This is an area with considerable research behind it (though it is not without controversy), so there is a great deal of information and evidence from which inferences may be drawn.

# CHAPTER FIFTEEN

# Obtaining Food

Having examined both the meals available to tyrannosaurs and their competitors for the available food, we can begin to integrate what we know from their anatomy and reconstruct what they were likely to have been doing ecologically and behaviourally. For the earlier dinosaur forms this information is rather more sketchy than for the later ones – we have far fewer fossils and far more competitors to confound the data – but for the tyrannosaurines things are much clearer. If, say, we find evidence of bite marks on a sauropod bone from a small theropod in the Late Jurassic, these may have come from an allosaur, ceratosaur or tyrannosaur, but unless there are some shed teeth near the bone, it is hard to come to an accurate conclusion as to the carnivore's identity. However, in the last parts of the Cretaceous especially, all carnivores larger than around 4 metres or so that had teeth were either tyrannosaurs or crocodilians, which are hard to confuse, so any bite mark or injury to potential prey can be much more easily assigned. Furthermore, the albertosaurs and tyrannosaurines were rather more specialised in their dentition and overall construction than were earlier forms, and were quite similar to one another. This means that it is relatively safe to take information we might have on one species and apply it to its close relatives, so while most of the knowledge we have comes from only a few species, it is probably applicable to them all. That said, more general areas of biology can be applied rather more widely across groups. Before exploring these issues further, however, we must address the great misnomer about *Tyrannosaurus* (also incidentally applicable to other tyrannosaurines and more distant relatives): was it a predator or a scavenger?

## Predation and scavenging

The first response to this should be that it is the wrong question to ask, because it implies a dichotomy in nature that essentially does not normally exist. Very few large carnivores can afford to be dedicated scavengers. It takes a rather special set of adaptations since it relies on waiting for animals to die or be killed, getting to them in time to make a meal, and being able to either wait for others to finish eating or to chase them away. Vultures can do this, soaring huge distances with very little energy expenditure and relying on scraps from kills, but no other large animal can. Certainly, many carnivores scavenge food, and even those that are highly specialised as predators opportunistically take dead meat if it is available (why pass up a meal?). The idea that lions are predators and hyenas are scavengers, for example, is incorrect, since both groups kill plenty of prey, as well as taking it off one another and other hunters when they can. Tyrannosaurs would seem to have been no different, so a better question is: where on the spectrum from predator to scavenger might they lie?

Some of the lines of evidence proposed for the dedicated scavenger hypothesis are clearly problematic, and may not even provide good evidence for a general ability to scavenge at all. *Tyrannosaurus* was described in this context as having 'beady eyes' by palaeontologist Jack Horner, but the truth is that *Tyrannosaurus* has some of the largest orbits of any terrestrial organism ever.[25] Small and beady they were not, though large and superb they probably were (and, of course, vultures have amazing eyesight). Similarly, the idea that *Tyrannosaurus* having an exceptional sense of smell made it a scavenger is flawed. Although such a sense would have helped the animals to track down carcasses, many dogs and sharks also have exceptional olfactory abilities, yet they are not pure scavengers.

Could a large tyrannosaur even function as a scavenger? Being the only large animal in its environments, a sole tyrannosaur could probably bully even a group of dromaeosaurs or troodontids to make them leave a kill, but would this occur often enough across the expanses of the environment to produce enough food for a multi-tonne animal? Few animals just drop dead, other predators do not routinely kill animals many times larger than they can eat, and both pterosaurs and birds would be able to travel much more rapidly to a corpse and required far less food. This would mean that an animal like *Tyrannosaurus* would be in a constant race to get to the limited available food, and even if it was an ectotherm it would require a large amount of food to keep going. It would also face serious problems from young members of its own species, which would be much quicker than it. Tyrannosaurs may have been efficient walkers and even runners, but this factor alone would not have keep them in meat. In short, the idea that large tyrannosaurs were purely scavengers is clearly problematic and there is no real data to support it – but did they get *any* sustenance from this method?

Some past studies have suggested that this was the case. There's a pelvis of *Triceratops* known with a major chunk missing and numerous bites from a *Tyrannosaurus*. It certainly demonstrated that at least one *Tyrannosaurus* fed heavily on the body, but this could simply have been late-stage carcass consumption.[26] Carnivores of all kinds tend to consume large prey items (those they can't just swallow in a few bites) in a stereotypical pattern, starting with the areas of major muscle mass (like the thighs) and continuing to the viscera, before moving to areas with rather less meat, such as the neck and feet, and typically finishing up with the head. Feeding by a large animal over a period of days, or feeding by a whole group of animals, can ultimately leave little of a carcass, but it is impossible to tell if the above example was a case of a tyrannosaur scavenging the last scraps from a long-dead ceratopsian, or from one that it had killed itself and fed on over a number of days.

However, there is now convincing evidence for scavenging as a result of unusual preservation of a fossil and the data it has provided. I was lucky enough to be able to describe the following example, which is based on a near-complete skeleton of the Mongolian hadrosaur *Saurolophus*.[27] The animal in question is beautifully preserved: almost the whole skeleton exists, the bones are in superb condition and there's even a patch of fossilised skin. About the only things missing are some bits of the tail and the hands, yet it is the condition of the humerus that is key: it is covered in bite marks. There are deep bites, and a huge number of parallel scrapes across the surface, especially on the area where the main muscles would lie, and they are of such size and depth that they could only have come from a large tyrannosaurine, and in this context it would have been *Tarbosaurus*. The contrast to the rest of the skeleton is remarkable. The humerus has hardly been destroyed, and much of it has been subjected to a great deal of surface damage: it looks as though someone has taken a particularly rough cheese grater to a large dinosaur bone. Parts of the humerus are undamaged and the rest of the specimen is in all but immaculate condition. I checked every other bone of the skeleton as far as I could, and found no trace of even a dent to any other element. So what happened here?

This pattern does not follow what we would expect from normal carcass-consumption patterns. Although there would have been some good muscles on the humerus, the femur and those huge leg muscles would surely have been the prime target, then the chest cavity with all its gory goodness inside, yet these remain untouched. This suggests that they were either gone already or inaccessible, and in either case implies scavenging. Wonderfully, the burial site of the specimen supports this: the animal shows evidence of having been transported to the site, where it came to rest on what was a sandbar or side of a river. In other words, it had died, moved, then been buried. Not only that, but the damaged humerus was lying uppermost and showed signs of erosion before it was buried and preserved, so it must have been exposed. It seems highly unlikely that a *Tarbosaurus* that would have been able to kill a large *Saurolophus* would then only have eaten part of its arm (spending quite some time doing it), before the corpse was somehow swept away from it downstream, and got buried in such a way that the only exposed part of it was the already damaged bone. It is much more likely that the dead animal came to rest buried on a sandbar, with only the uppermost parts exposed, and that these were then discovered and consumed by a tyrannosaur. If this is what happened, it was clearly a case of scavenging, and the individual *Tarbosaurus* in question did not kill the animal it was feeding on. We return to this example later as it also contains useful information on tyrannosaurine feeding patterns, but it does provide secure evidence that tyrannosaurs were scavengers on at least one occasion.

Setting this aside we can now turn to predation, and here again the fossils provide convincing evidence that it occurred. In this case we move to North America and the activities of *Tyrannosaurus*, though again the key specimen is a hadrosaur. In 2013 a specimen was described with an injured tail. Tail injuries are pretty common in hadrosaurs, and they seemed to have suffered a disproportionate number of injuries to this part of the anatomy, but in this instance there was a massive ball of bone at the juncture between two centra that had fused them together. A scan of these elements revealed the presence of a foreign body in the form of a *Tyrannosaurus* tooth, which caused the injury and the resultant mess. The tooth could hardly have been wedged in there by accident, and the bones had actually grown over it to totally bury it inside. The obvious conclusion is that the tooth became lodged in the hadrosaur as the result of a bite to the tail.[28] The animal not only escaped, but survived long enough for the wound to fester and produce the major deformation, which may have occurred months or even years after the event. The most plausible explanation for such an event is a deliberate strike by a tyrannosaur, and while obviously this time the intended meal escaped, it was certainly a predation attempt. Other injuries to hadrosaurs show similarities: in another hadrosaur from the US, a neural spine is missing from a caudal that appears to have been bitten through, and yet another has a tyrannosaur tooth wedged into one of the bones of the leg – though it is not clear if this shows healing and whether the bite could have been delivered post-mortem.

So tyrannosaurs (or at least the tyrannosaurines) clearly did both scavenge and actively predate on other animals. Given the overall limitations to scavenging, it seems likely that the majority of tyrannosaurs were primarily predatory and relied on themselves for obtaining their meals. Scavenging was very much an option, though, and one taken advantage of when possible, but it probably did not provide the majority of their food intake.

## Who is on the menu?

In terms of the species fed upon, almost anything a large tyrannosaur came across could at least potentially have been eaten, and in most environments that we know of, that's potentially quite a large selection of animals (Fig. 16). Even so, the larger tyrannosaurs clearly favoured hadrosaurs. Unlike ceratopsians, these were often ill equipped to fight back even in a crisis, so this is no great surprise. A major survey of elements of both groups from the productive Late Cretaceous fossil beds of Alberta, Canada, revealed that around 14 per cent of recovered hadrosaur bones showed traces of feeding from tyrannosaurs, compared with around 4 per cent for ceratopsians.[29] However, this information may not actually reveal a genuine preference, but rather that the more common hadrosaurs were inevitably being targeted more often than other animals. To date, there are no records of damage to either ankylosaurs or sauropods in tyrannosaur-dominated faunas, but both of these were rather rare, so this is probably due more to a lack of data than their being overlooked as prey items. Gut contents of tyrannosaurs and their coprolites also suggest the consumption of various hadrosaurs.

Most impressive of all is an indication that these animals may have been cannibalistic on occasion. The evidence for craniofacial biting and intraspecific combat is strong, but a *Tyrannosaurus* metatarsal also shows evidence of bite marks. While it is not impossible that the bite was delivered during a fight, it seems rather more likely that it was made during

*Fig. 16 Representatives of dinosaurs living alongside tyrannosaurs: here a fauna representing the Lance Formation from the end Cretaceous of North America. Clockwise from top right: a pachycephalosaur, ankylosaur, early bird, troodontid, ornithomimosaur, sauropod, ceratopsian, hadrosaur and oviraptorosaur. All of these species (and indeed many others such as crocodiles, pterosaurs and champsosaurs) are known from this one time and place.*

feeding (and indeed, the pattern of the bite marks better matches this supposition). Similarly, a *Daspletosaurus* in Alberta has multiple marks on the skull, the pattern of which indicates that they must have been made post-mortem, suggesting that another tyrannosaur, and quite possibly another *Daspletosaurus*, fed upon the animal. It is unlikely that tyrannosaurs made a major habit of hunting and killing each other, it would have been extraordinarily dangerous to do so habitually, and such prey would be rare and hard to find, so this is much more likely to have been a case of scavenging. Clearly, the diet was varied – though it is notable that for all the numerous feeding traces, there are just two incidences where stomach contents have been found. So what happened to what they were eating?

Much of the available evidence points towards carnivorous theropods in general targeting juvenile animals. This is pretty much a universal pattern for carnivores, as juveniles are small, lack the experience and weapons of adults, and have to forage for longer periods in suboptimal conditions, making them ripe targets. Few carnivores attack animals larger than themselves, with most taking prey a fraction of their size (usually less than 50 per cent – the exception being group hunters), and of course adult hadrosaurs and other dinosaurs were of similar mass or even larger than many contemporary predators. Dinosaur juveniles were also numerous: most large mammalian herbivores have just one or two young a year, but dinosaurs laid nests containing dozens of eggs, perhaps even several times a year. There are biases in the fossil record against small and juvenile animals, but the profound lack of juveniles is notable, and we know that in modern ecosystems for almost all studied species, the single biggest mortality factor for juvenile animals is death from predators. It is notable that the one tyrannosaur coprolite found represents consumption of a juvenile hadrosaur,[30] that the record of stomach contents in a *Tyrannosaurus* includes elements of a juvenile, and that all three fossil specimens showing bites and survival are those of juveniles.

Adults, then, were generally to be avoided; even a large hadrosaur might have been capable of injuring a tyrannosaur if attacked, and with sauropods being huge, ankylosaurs being armoured and some ceratopsians being well armed, juveniles do appear to have been the prime meal available, at least when it came to active hunting. However, just as juvenile herbivores make mistakes and are naive about predators, so too young predators can make errors and take on targets that are well out of their league. Plenty of nature documentaries show young lions trying to tackle an adult buffalo, or struggling to get to grips with a porcupine, and it is not hard to imagine a young tyrannosaur mistakenly attacking an adult ankylosaur or ceratopsian, and paying for that error. It can't be proved, but I suspect that the incident of a tyrannosaur having bitten on the horn of a *Triceratops* (mentioned earlier) may well have been one such case. It's difficult to conceive of an experienced hunter being caught out and standing face to face with such a well-equipped adversary, and moreover going so far as to try and bite the horn. This is not the only available explanation, but I think it has some merit.

## Finding a meal

Having discussed the issue of what animals the tyrannosaurs targeted, we now come to the question of how they did it. With their excellent vision and sense of smell, prey should not have been hard to find for the average tyrants. Whether looking for individuals or herds of live animals, or potential meals in dead ones, smell would have helped them to track things beyond the line of sight, with vision providing details closer in. However, closing in to make a kill may have been difficult. Almost all modern predators hunt primarily through one of two basic strategies, ambush or pursuit. In the case of the former, they can lie in wait for prey to come near (or approach it carefully and undetected), then strike from a short range to take it down. This strategy encompasses classic sit-and-wait hunters like crocodiles and preying mantids, and also things like cheetahs. If the prey has too much notice, it can increase the gap between them, or the element of surprise is lost and the strike fails. Group hunting can be effective in this strategy, as is the case in hunting by lions: during the course of an ambush, no matter which way the prey goes there's likely to be a predator within range to make an attack. Alternatively, a predator can approach prey rather more boldly, and simply try to outrun it over distance, wearing it out and exhausting it before it is overwhelmed. This is a common approach used by canids such as wolves and hunting dogs, and also hyenas. Cooperation helps here, too, as the hunting animals can take turns in harassing a herd or individuals, so that none of them is too tired, while the prey is under constant pressure.

Tyrannosaurs were relatively fast, at least this was the case in the smaller species, and the younger individuals of the giant tyrannosaurids. A quick sprint may well have been used to catch out an off-guard herbivore or small animal, and get in close enough to make an attack. Getting within range through stalking may have been tricky for the small tyrannosaurs, and almost impossible for large tyrannosaurines. An ambush only really works if you can narrow down the distance to the target so that you can reach it before it can react effectively. You might be quite a bit faster than your intended dinner, but you will tire eventually, and if you have to cover a good distance to get to where the prey started because it had a big head start, then you will probably tire before you reach it. In the modern world even lions can struggle to approach prey in the African grasslands; that classic cat-like crouch and crawl, shoulders rippling, is utilised to minimise their profile and help them approach wary prey animals, which can often see well over the grass as they stand tall. Lions living in open country tend to hunt more at night, when the lack of cover is less of an issue. If a crouched lion might struggle to hide, how exactly would an adult *Tyrannosaurus* have concealed itself even if the plants were a metre or two high?

The short answer is that it could not possibly have done so, so any attack would have had to be from pretty obvious cover (such as a large stand of trees), or it would not have been a short-range ambush, but a long-distance pursuit. A better understanding of the environments the tyrannosaurs tended to favour would help answer this question; it is notable that forests abounded in some areas at the time of the large tyrannosaurs, and some evidence suggests that this is where hadrosaurs preferred to feed, while ceratopsians stayed in the open. Perhaps adult animals shunned the deep forests as they were hard to move through and left them vulnerable, while tyrannosaurs might have kept to the trees where they could strike at unwary juveniles. It's speculation, certainly, but bounded by the obvious problems of hiding a 12 metre-long and 4 metre-tall biped.

A longer term and more obvious approach might have been the way forwards for the giants. You don't have to be that fast to easily catch a hadrosaur, as long as you can keep up the pressure by not falling too far behind; you might be able to overhaul it, or a charge into a herd could easily reveal an animal that was injured, ill or old, or perhaps you could snag a juvenile that had panicked or made an error in running too late or the wrong way. Tyrannosaurs were certainly efficient runners, and this at least is tentative evidence to support this idea: speed might have been much less important that endurance. Since I wrote these last couple of paragraphs a paper has come out suggesting that hadrosaurs were generally faster than tyrannosaurs over short distances,[31] which does support this idea. If hadrosaurs were relatively fast, then a short-range ambush would have been pretty worthless, so a long-distance pursuit would have been the only real option (though juvenile hadrosaurs might have been slower than the adults, and of course potentially more slow to react).

## In for the kill

When it comes to the way a tyrannosaur dispatched its prey, a number of different ideas have been put forwards for the manner in which it would strike at prey, and attempt to disable or kill it. In the case of smaller animals, there was probably only one real option: a single large bite would cripple or kill them almost instantly. However, larger prey would require something more; for all the talk of tyrannosaurs targeting juveniles, it must be borne in mind that even a half-grown hadrosaur might have been an animal of 4–5 metres long and weighing half a tonne. That's small compared with most adult tyrannosaurids, but it was still a hefty animal and not something that could be easily overpowered in an instant. Predators do take risks whenever they hunt prey, and a single injury can ultimately be the death of them (hence in part the preference for small and weak prey), so battling with an animal for several minutes is not ideal, and a quick, or at least safe, dispatch is to be preferred.

Dinosaurs did have 'weak points' that could have been exploited. The most obvious was the neck, where major arteries and the windpipe lay, and any serious bite here was likely to have been fatal pretty quickly through blood loss, suffocation or both (big cats typically kill through suffocation). In the case of animals like ornithomimosaurs and therizinosaurs, their thin necks would probably have been severed by a large bite, causing instant death. With the ornithischians things would have been trickier, since most of their vulnerable tissues were on the underside of the neck, and as a result rather harder to strike at since the animals were low slung. If the prey animal was large the neck would have been big, and it could have been hard for a tyrannosaur to get its mouth around the key area, but if the prey was small, the predator could bite down and the cervical vertebrae would be in the way. These could have been bitten through (thus severing the spinal cord and making for an instant kill), but even with the bite power of a large tyrannosaur this may not have been easy to accomplish, and in the case of ceratopsians and ankylosaurs, the frill and armour respectively would have been awkward obstacles at the very least. More obvious is the fact that in most cases the prey would probably be actively running away from its aggressor, and as a result the neck would represent a more distant target than other parts. Why catch up to the whole tail and body, then run past to bite at the neck? Doing this might represent a lot of extra effort to engage a difficult target, when a much more inviting option was nearby, namely the base of the tail.

The huge caudofemoralis muscles ran from the femur and up most of the first third of the tail in all dinosaurs. The tail gave a huge amount of power to the legs for running, was full of major blood vessels and was not surrounded by bone. A heavy bite anywhere around the thigh or first part of the tail might well have crippled an animal, leaving it unable to run and bleeding badly. Also, the tail was one of the first things a pursuing hunter would encounter in a fleeing animal, so using this technique would have reduced the chase distance, which would have been important especially if the prey animal was fundamentally faster than the tyrannosaur. Notably, there are two fossil hadrosaurs showing injuries to the tails from tyrannosaur bites, and another with a wound to the leg; it's a very limited data set, but it does nonetheless point to this as a strategy.

A bite from a large tyrannosaurid would potentially have been devastating. The power behind it would have enabled it to go deep into muscle, and maybe even break through some bones (as seen in the crunched *Triceratops* pelvis). The serrations on the teeth would also have allowed the teeth to cut into and through tissues. More damage probably came from any actions as a result of the bite. Analytical models and other such studies understandably tend to focus on static biting – a simple interaction between the teeth and skull and a static object like a bone – but in reality the prey would probably have been struggling, making things worse, and the muscular neck of the tyrannosaur would have had the potential to rip out a chunk of whatever it had in its mouth.

The bite would have been effective, and given the much-reduced arms, it would have been more the primary weapon to dispatch prey than any other part of the anatomy. However, it has been suggested that both the arms and the legs would also have been important, and regularly employed even by the tyrannosaurines when hunting. It can certainly be imagined that a mammal or lizard may well have been pinned with a foot by a small tyrannosauroid or a young individual of a larger tyrannosaurid, and that similarly, an adult tyrannosaurine might simply have all but stepped on a hatchling-sized dinosaur as an alternative to reaching down to bite at it. The obvious problem with employing the legs is that tyrannosaurs are bipedal, and using one leg to try and strike at, or even hold down, a desperate and struggling animal could well have led to the tyrannosaur being toppled over. Surely kicking out at even a half-grown ceratopsian would have been a bad idea. The legs of a tyrannosaur were positioned under the body, so the prey would not just have needed to be caught, but almost overrun, before the legs could be employed, and while they were obviously powerful and capable of giving a strong kick, they wouldn't have had a huge range of motion (especially laterally), and quadrupeds at least, with their four legs and low centre of gravity, would have been hard to topple or pin down. The legs may indeed have been brought in to help subdue prey on occasion, but it is doubtful if they would have been used regularly for this purpose – even as an adjunct to the mouth – when dealing with large prey.

Still more surprisingly, it has been suggested that the hands were integral to predation in the tyrannosaurines. The proportionally longer arms of the early tyrannosauroids and their kin may have been employed on occasion – they were relatively long, strong and had strong grasping fingers and curved claws – but would the tiny arms of the giants have been of any use? Certainly, they were not exactly weak, despite their size, but it is hard to imagine that they could have had a major role in predation. After all, engaging the arms could only occur once prey had been almost overtaken, which would seem impractical at best, and their position under the body was hardly suitable for grappling with an animal even if they could reach down and get a good grip. Foremost, though, is again the point that tyrannosaurs, like other theropods, were probably mostly hunting prey much smaller than themselves, and the arms would simply not reach down far enough and be of little use for getting at a small animal. A more extreme suggestion is that giant tyrannosaurs would knock over adult ceratopsians, using their arms in part as leverage. Again, while I'd say this isn't impossible, it would have relied on them attacking large, heavy and dangerous animals, and catching them fairly precisely midway (which would seem especially hard against a manoeuverable animal that would probably run away or face its adversary) with their arms, and only then trying to bite them.

## Table manners

To discuss how prey animals were actually eaten by the tyrants, we return to the scavenged *Saurolophus* specimen mentioned above. This is informative not just because it shows evidence of scavenging, but also because the pattern of the bites on the humerus provides details about how a large tyrannosaurine applied its mouth to its food. At the ends of the hadrosaur humerus there are deep punctures: essentially pits that represent the teeth of the jaws being driven into the bone. These obviously represent a strong bite (to make holes in solid bone), but also align on the front and back of the bone, implying a bite with both sets of teeth together. The bites on the rest of the humerus show a very different pattern.

The humeri of hadrosaurs had a very long and flat crest to support the muscles, and this is where the vast majority of the feeding traces lie. These are scrapes across the surface of the bone that are typically long, and occasionally quite deep grooves. The marks are parallel to one another and represent the premaxillary teeth repeatedly hitting the bone, then being dragged across it. These must be from the premaxillary teeth and not those of the dentaries, given how close they are to one another at the starting point and the number of them. On the reverse of the crest lie still more of these marks in the same pattern, but they do not align with those on the upper face. In other words, these marks were also made by the premaxillary teeth, not the teeth of the upper and lower jaws working together: the *Tarbosaurus* was dragging its teeth across the surface.

Tyrannosaurines in particular have in the past been characterised as crude feeders, simply using jaw power to bust into or through bones, then swallowing whatever piece came off as would some giant crocodile. The truth is that they were certainly capable of a much more measured approach. Deep bites at the joints of bones would have helped sever the ligaments and tendons holding the skeleton and muscles together, and freed up parts to make them easier to manipulate. Then the premaxillary teeth were used to scrape off the major muscles for consumption; it's notable that most of the major muscles of the upper arm attached to the crest, so the pattern of bites concentrated here matches what we would expect of the best places to feed. The tyrannosaur even seems to have flipped over the bone to feed on the other side, rather than simply applying the dentary teeth to the initial bites, it was clearly adapting its bite power and style to best suit the shape and condition of the meal. With the limb perhaps severed from the torso, it would also probably have skittered around on the ground, and it is likely that it was pinned down with the foot to hold it in place during feeding. Interestingly, a recent study showed that tyrannosaurines probably fed in a rather raptor-like manner, using the head and neck in line with the body, and pulling back up when biting (just like a hawk or owl), not twisting like a crocodilian.[32] Their musculature was arranged in a very similar manner to that of birds, which in this case offer a superb model for how they might have stripped flesh from a meal.

This is not to rule out other activities. Specimens like the bitten *Triceratops* pelvis show numerous deep bites and even chunks of bone having been removed; bones contain nutrition in the form of marrow, as well as calcium and other minerals, and they would have been attacked and consumed on occasion. Bite marks on herbivore bones are certainly more common in tyrannosaur-dominated faunas than in those where allosaurs or megalosaurs dominated, and this suggests that the tyrannosaurs were biting bones with more frequency, or perhaps that they were less concerned about biting into bone, having stronger skulls and teeth suited for the task. Still, this doesn't mean that they crunched their way through all and sundry when tackling a decent-sized carcass. They may, however, have done so when feeding on small prey. It would be difficult for a large tyrannosaur to strip the flesh from prey that was only a small fraction of its size (perhaps even a fraction of the size of its head), and if it was biting on small bones the carnivore would be rather less likely to break or shed teeth. In these cases the meal would probably have been broken up into a few bites and swallowed. Intriguingly, the *Tyrannosaurus* coprolite containing small hadrosaur bones shows that these have largely been smashed apart. The digestive tract of a large tyrannosaur might well have been a rather hostile environment – the stomach is, after all, a big, churning sack full of acid – but this evidence does suggest some degree of oral processing before the hardrosaur was swallowed. It implies that the *Tyrannosaurus* bit the body multiple times and macerated the pieces before consumption, perhaps to aid digestion.

# CHAPTER SIXTEEN

# Behaviour and Ecology

There is a lot more to what we understand about dinosaur behaviour and ecology than what has been covered so far. The context of other species and their influences, and the fundamental limitations of the lifestyles of tyrannosaurs, do go a long way to providing a framework to understanding how they lived, but we can try to go beyond this, and in some cases there are important hypotheses about tyrannosaur biology that we can work from. Many have little or no support in the fossil record, but they are viable possibilities nonetheless and are at least worthy of discussion.

An important hypothesis is that tyrannosaurs may have formed social groups, and in particular that this behaviour may have been linked to group hunting. This has specifically been suggested for the albertosaurine *Albertosaurus*,[33] but inevitably in some quarters it has been taken to apply to perhaps all tyrannosaurs. Determining social behaviour in any fossils animal is especially hard, and there has been a major issue in the past with palaeontologists conflating a variety of animal interactions with them being social.

The term 'social' is often applied to animals that are living together, but really it should have a much more limited scope. Although definitions vary between ethologists (those who study animal behaviour), something closer to 'spending time with one another and interacting' would be more accurate. Almost any interaction between two individuals of a species might be deemed social under this concept, and while obviously two animals fighting are having an interaction of some description, they are not necessarily living together. In short, the terminology is already awkward enough without things being further confused. So animals might live together and be social (think gorillas, or meerkats, or at extreme levels ants and bees), or be primarily solitary and live alone (such as tigers or rhinos – except when breeding, of course), but as ever these are only two ends of a spectrum. Plenty of animals hang around in groups without necessarily having any kind of social interaction. Crocodilians are generally quite tolerant of one another, but they do not form social groups that involve well-structured hierarchies, numerous interactions, social bonds, mutual activities such as grooming, and shared responsibilites like watching for predators, or even sharing food. None is a real threat to any of the others, but they are not like packs of wolves.

## Living on one's own

Animals are also rather plastic when it comes to these relationships. Most bears are normally solitary, but when the salmon are spawning high concentrations of grizzlies congregate together – yet again, though, this is an aggregation, and does not represent sociality. Lions are often considered highly social animals, but males may live alone or in pairs for extended periods in their lives, and can survive and even thrive: sociality is not essential to them. Cheetahs are an exceptional example because of the ranges and complexities seen in their interactions. Typically, cubs leave the mother as something of a group, with females later becoming solitary and any brothers (normally two or three) living together in a group, so there are both social and solitary members divided by sex. However, males may live alone, and when breeding females may join male groups for extended periods, providing exceptions to both norms, so a single animal may alternate many times in its life between the two.

Evidence for sociality in dinosaurs is really rather limited. There are plenty of trackways that show animals of probably the same species travelling in the same direction, and whole groups were trapped in sandstorms, mud pits and other disasters, but none of this evidence points to sociality. A flash flood could come down and bury a group of grizzly bears feeding together, but they are not social or even normally aggregate, and even solitary species may come together if they have to migrate from and to the same general points to get to their food, or to breed. What these patterns tend to do is provide evidence for aggregations, not sociality. A few, perhaps even most, dinosaur species probably *did* exhibit degrees of social behaviour and interactions, but can we separate them out from the others based on the available data?

Even animals that lived in groups might easily be preserved as single individuals in the fossil record; accidents, diseases, predation and the like generally take out only one individual on a given day, preserving just a single specimen. But if these animals regularly spent much of their time in groups, then those disasters might well preserve whole herds, packs or families. Various hadrosaurs and ceratopsians in particular (and indeed some specific species) seem to be regularly found in groups, and evidence from trackways and nesting sites supports the idea that they were often present in large numbers in small areas. Yet even though they were found in groups, this does not necessarily mean that they were social animals.

It is notable that features used in sexual displays, and especially those under mutual sexual selection, often also function as social signals of status. In other words, they can be used equally to attract a mate and to demonstrate that an individual is fit and healthy, and a senior member of the herd. We do have tentative evidence in some species for social structures and interactions, which suggests that dinosaurs did not just aggregate, but may also have had more complex behavioural interactions.

In the case of tyrannosaurs, bosses and hornlets over the eyes of many (and more elaborate crests in the basal tyrannosauroids) give some support to this idea. However, much better information comes from the apparent damage left on the skulls of a number of tyrannosaurs. As noted before, bite marks on dinosaur bones are relatively rare, though they are apparently comparatively common where the larger tyrannosaurs existed. However, carcass-consumption patterns suggest that skulls would not normally be fed on (or at least would be targeted last), because they contain a lot of bone and not much flesh, and this is doubly true of the snout compared to, say, the back of the head. Why, then, do a number of tyrannosaurs (and for that matter at least one other large theropod, the Middle Jurassic *Sinraptor*) have evidence for not just bites, but often healed bites around the tip of the snout and the anterior part of the dentary?

The interpretation of these marks is that they were the result of combat between tyrannosaurs of the same species.[34] Two animals squaring up to one another would try to bite each other, and the head is a rather obvious target, and perhaps also a place when you can minimise damage. It may seem an odd choice, but due to the strength of the tyrannosaur skull, combined with the thick bone on the nasals, hornlets on the eyes and any extra keratin, the head might have been a relatively safe place to be bitten, whereas a bite to the neck or body could easily be fatal. The traces when they appear certainly don't look like feeding bites (signs of healing in some invalidate that idea as well), and few predators could make much of a living taking on equally sized and equipped animals as meals very often, so such scenarios wouldn't explain why these marks occur so often.

There does therefore appear to be a strong suggestion that the tyrannosaurs engaged in combat, but this would probably have been merely one aspect of their social interactions. Animals often fight, even to the death, but they generally don't take on fights they think they will lose. Most go through various rituals of ramping up aggression, both to indicate a willingness to fight and to size up the opposition. We're all familiar with dogs raising hackles and baring teeth while growling, and many animals have similar routines: various antelope, for example, walk alongside each other to try and estimate the size of the opponent, and many engage in some light sparring. You want to come out on top, but it's not worth it if there's a good chance that you could lose. For the tyrannosaurs, the evidence does suggest that, although these interactions could turn bloody or even fatal, there might have been a lot more biting than serious wounds being delivered. Although proving such behaviour based on scanty data is difficult, the evidence is certainly consistent and a reasonable interpretation can be derived from it.

Craniofacial biting and sociosexual crests might have been used to signal to other members of the species and could thus be considered 'social' signals, but this does not necessarily indicate sociality. Animals that came together only to breed, or that were antagonistic over things like territory or a kill, would still benefit from these features and an ability to advertise themselves, but they would certainly get more use from them in a social hierarchy, which leads us to *Albertosaurus*. As already stated, large carnivores are generally rare, and mass mortality events that would kill a whole group of animals are also uncommon, so even if large theropods did often or even habitually live in groups, finding evidence for this might be difficult. However, one quarry in Canada provides strong evidence of exactly that.

## You don't win friends with salad

Back in 1997, after a great deal of detective work, an old quarry originally opened up in 1910 by Barnum Brown, an American palaeontologist, was rediscovered by a team of palaeontologists from the Royal Tyrrell Museum of Palaeontology in Alberta. This purportedly contained numerous tyrannosaur specimens, so a great deal of effort was put into its rediscovery and new excavations. Huge numbers of remains were uncovered, and while the animals were mostly disarticulated, enough was collected to determine that there were more than 20 individuals present, and of a range of sizes from mid-sized juveniles through to adults.[35] This does appear to represent a genuine aggregation that sampled a real group, and not a result of some other event that could have grabbed a disproportionate number of meat eaters.

Carnivores can be stranded together when caught in 'predator traps': something like an animal trapped in deep mud might draw in a predator to what seems like an easy kill, only for it too to get stuck, and draw in another animal, and so on. Similarly, a rotting carcass stricken with some brutal bacteria might well bring in a carnivore, then kill it rapidly (people can die very quickly from food poisoning and this does happen to wild animals, too); again, the apparent supply of bodies at such a scenario can relate to scavengers that are drawn in to the dead animals. However, you typically find the odd herbivore in both cases (it usually started off the chain), and multiple species of carnivore are drawn in, not just one. The *Albertosaurus* group, then, appears to be a genuine one, though again we should be cautious about extending this to be generally true of this genus, let alone tyrannosaurs collectively.

At least one tyrannosaur therefore may have spent some time in a group, and even used social interactions and intraspecific combat as a dominance signal. Assuming that this was normal for *Albertosaurus*, how might it have influenced its behaviour? Having a number of large predators form a group would increase the amount of food they would need, but also increase their capability to obtain it. Packs of animals can tackle species or individuals far outside their normal range of prey size, so this could effectively have allowed the group to tackle adult hadrosaurs and ceratopsians.

Although they don't exhibit any form of specific cooperation, aggregations of crocodilians do form groups of a kind and hunt together for mutual benefit. When fish are travelling en masse, caiman spread themselves out across rivers. If they each simply patrolled a section of river, the fish would find them relatively easy to avoid, but if they are spread out in a line across the whole width of the river, the fish have nowhere to go and pretty much every animal is guaranteed to catch something. So here there is a situation where there's no real coordination (compare this to how lions hunt in an actual ambush, with some driving the prey towards others, or hyenas taking turns harassing the prey they are chasing), but everyone does better from the system than if they went at it alone. However, tyrannosaur expert Phil Currie, who has studied the *Albertosaurus* remains, has suggested a much more radical concept: namely that not only was this a structured group that hunted together, but that the integration of both adults and juveniles was key.

Currie's idea is that the longer-legged and faster juveniles would act to round up and direct prey, while the larger and more powerful adults would close to deliver a killing blow. This is certainly not impossible, but I think it's rather unlikely. Assuming this was the prime or only mechanism for getting food (as it implies that either age group on its own was poorly equipped to tackle potential prey), both sides would appear to risk being stuck. As we have seen, infant mortality was huge for young dinosaurs, and even with the extremely rapid growth of the young, they would not have reached adult or near adult size for quite some years. If Currie's interpretation is correct, the adults would have had to keep a production line of babies going in order to have a constant supply of animals of the right age (too young and they wouldn't have been of much use, and too old and they'd have been slow and competitors, not cooperators). A few bad years of drought or poor breeding conditions would have resulted in adults being left without this age group, and would have left them high and dry, too. While obviously such information is also rare, we have yet to see much in the way of tracks, feeding traces or stomach contents that would support the animals working in this way, or targeting larger prey. The idea that the *Albertosaurus* find represents a genuine aggregation is a strong one, and the animals may even have been truly social – but such a specific idea about exact hunting patterns needs some strong evidence to support it, and for me it falls rather short at the moment.

Similarly, other ideas are hard to assess without evidence, but are at least plausible. For species living in higher latitudes, winter would have brought cold, especially in the Late Cretaceous. Places like Canada and Alaska (while warmer than they are today) would have seen freezing temperatures and snow in winter, and food would have been scarce or animals would have had to migrate to greener pastures. Tyrannosaurs had two options, to stay and wait out the conditions, relying on fat reserves and perhaps taking some animals that stayed behind, or to move with the herds and migrate themselves. The efficient gait of tyrannosaurs may have facilitated the latter, but this idea is currently unproven.

Even if tyrannosaurs were non-migratory, their prey would probably have often been on the move: large herbivores can eat themselves out of food quickly and may be forced to move around, and hadrosaurs at least sometimes lived in mixed herds, so juveniles would be with large adults. Tyrannosaurs may have established territories and patrolled them, waiting for prey to move through or into the area, or they may have been closer to nomads, following prey and constantly looking out for stragglers or opportunities. Large animals by necessity tend to cover large areas (either within a large territory or by moving regularly), and also therefore tend to encounter a wide variety of environments; certainly, populations as a whole would struggle to stick to, say, just forests or just estuarine habits. Smaller animals occupy environments rather differently, and earlier tyrannosaurs may well have been specialists in this sense and stuck primarily to one area.

Some aspects of tyrannosaur ecology are reasonable to infer based on what we know of the ecological dynamics of large carnivores. Most illustrations of tyrannosaurs show them hunting (generally some giant and well-armed and therefore thoroughly unsuitable prey like a large *Triceratops*), eating or fighting another tyrannosaur. Almost none of them show what probably took up most of their day: sleeping. Doing very little is a pretty common feature of the lives of large carnivores. Unlike in herbivores, their food is highly nutritious and easy to digest, and when they do make a kill or come across a carcass there is plenty to eat that might keep them going for several days. They are not looking for food 24/7, and there's not a huge amount to do in between doing so, and little point in exerting energy unnecessarily. A major chase or attack can be exhausting, and despite what the nature documentaries show, even the most effective and efficient hunters are only successful around a third of the time. Many attacks therefore resulted in little more than a tired tyrannosaur, so long periods of rest or even sleep would have been likely. If the animals were hunting primarily at night or in low-light conditions, they would sleep for most of the day, too.

It is possible that tyrannosaurs were nocturnal or crepuscular (operating primarily at dawn and dusk). As discussed in the previous chapter, large tyrannosaurs would have struggled to hide given their size, but operating in low light conditions would have helped them do this considerably. It's quite hard to see even a large predator in poor light, and tyrannosaurs may well have taken advantage of reduced visibility at night. This is naturally very hard to analyse, though some preliminary work has been done to look at eye shape in dinosaurs and pterosaurs to determine their activity cycles.[36] Although the results of this work are rather uncertain, we do know that in general large eyes correlate well with visual acuity, and can help with vision in low light. The later tyrannosaurs did have huge eyes, so might well have been able to see relatively well in low light, and perhaps better than their prey species, so this is a possibility that should not be overlooked. Being largely nocturnal may additionally have helped prevent such big animals from overheating during strenuous activity like hunting, so could also fit this hypothesis.

**Get on down**

One relatively little-studied area of tyrannosaur biology is how they might have slept, and more importantly for the giants in particular, how they might have stood up. There are some wonderful tracks for smaller theropods which show that they at least sometimes crouched down on all fours, or even all fives, with the hands, feet (including the metatarsals) and then the pubis all resting on the ground. Others were preserved in a similar posture, with the head tucked alongside or under one arm, and even with the tail wrapped around the body. Such postures would have been fine for something like *Guanlong* or *Eotyrannus*, but what about the giants? They were so much heavier, and especially given those small arms, would such postures not have placed excessive weight on the pubic boot, and even on the ribs and chest when trying to breathe? The extra-large and robust gastralia might have helped to support the chest generally, but could also have helped to stop the whole thing collapsing if the animal lay on its front. Still, the fact that this crouched posture was clearly utilised by both very early and much later theropods of all kinds (and is almost uniform in birds) suggests that it might have been adopted by the tyrannosaurs. Rolling onto a side after crouching to get down to the ground might have been an option, too, though it is notable that while even huge animals like rhinos and elephants do lie on their sides, a more normal posture for them is to lie with the legs tucked under the body. It is relatively easy to stand up quickly from this kind of position if danger suddenly threatens, and a large tyrannosaur on its side would have needed to roll back onto its front to have much hope of standing up quickly.

Actually, standing itself would have been a challenge for the big species: the large head resulted in there being quite a lot of mass upfront, and the small arms, even if well muscled, were hardly well placed to provide a great deal of leverage. Large animals can appear quite ungainly when standing[*] and even more moderate animals like horses and cows either stand up slowly, or require quite a bit of twisting and sudden jerks to get upright. In the case of the big tyrannosaurs, the arms would have provided some thrust to get the torso clear of the ground, and perhaps the head was flicked back and up with the powerful neck muscles as the legs extended, rocking the animal back, then a step forwards would get some support at the front as it moved into a standing position.

From here, we could move into the realm of real speculation; there are plenty of common behaviours in extant animals (including birds and crocodilians), which we simply cannot assess in tyrannosaurs. Due to a lack of data, or even the abilities to make reasonable inferences (as in the case of parental care discussed in earlier chapters), they remain unknown, and in some instances unknowable. The possibilities

[*]Watching a giraffe get up in a hurry can be quite an education.

for these behaviours are mentioned here not to suggest that any of them occurred in tyrannosaurs, but to demonstrate what else we may yet discover. Quite often incredible evidence comes to light which shows that dinosaurs (and other extinct species) had certain traits or behaviours, and these are regularly treated as absolutely stunning revelations by the media and public alike. However, what is so remarkable about them is not the behaviours themselves, but that we are able to demonstrate that they occurred. Brooding is a case in point, as its occurrence seemed quite obvious and even extremely probable to researchers, but discoveries of fossil specimens of oviraptorosaurs on nests were seen as amazing not because of the idea that dinosaurs sat on their nests, but rather because this could be unambiguously demonstrated. What, then, might we consider at least possible for tyrannosaurs?

Pair bonding is common in birds, with males and females staying together for years at a time, perhaps maintaining a territory and working together to hunt as well as to rear their offspring. Such links often require positive reinforcement, and there may be elaborate courtship rituals and activities such as mutual feeding to help support the bond (albatrosses, for example, go through some beautiful rituals of this kind). Actually acquiring a mate may require some feat of combat, involving not just males fighting other males to mate with females, but perhaps the reverse (this occurs in some living species), or even the female requiring to be effectively subdued by a male to demonstrate his virility. Many birds also engage in 'helper at the nest' activities: in these cases, juveniles from previous years that are not yet ready to breed or cannot find a nest help parents rear their siblings in the next breeding season, and mammals that live in groups may cooperate to raise offspring, too. Complex hunting behaviours are not impossible (just undemonstrated), with, for example, crocodiles recently having been suggested to be tool users: they've been seen collecting twigs that birds need to find to make nests, then striking at the birds when they come to collect the twigs. None of this behaviour is necessarily even likely in tyrannosaurs, but should evidence ever be found for anything like it in them (or in other fossil groups), I would not be surprised.

Collectively, then, we know quite a bit about the behaviour and ecology of the tyrannosaurs. Classic controversies such as predation versus scavenging have been dealt with, and we have information on what they ate and how. While there may be no direct evidence about hunting, we can make some convincing predictions about how tyrannosaurs may have hunted, and the approaches they must have taken given the size of the larger forms and the lack of cover available. Behaviour and ecology are difficult to deal with in fossil forms since so little information is generally preserved, but we can get solid ideas about these aspects of the tyrannosaurs by bringing in comparisons with living species.

PART FOUR
# MOVING FORWARDS

# CHAPTER SEVENTEEN

# *Tyrannosaurus* Fact and Fiction

All too often where tyrannosaurs – or indeed any dinosaurs – are concerned, the spotlight falls upon *Tyrannosaurus* alone. Of the dinosaurs as a whole it is by far the best known by the general public, and as a result pretty much every new discovery of a dinosaur (and even many non-dinosaurs) seems to get compared to it. Such is the cachet and recognisability of the tyrant king, that it has become a touchstone for the media whether or not it is relevant to the story at hand.

*Tyrannosaurus* was certainly a wonderfully interesting animal in its own right, but the overly intensive attention to it as some kind of baseline comparison is often misguided. It was no more a 'typical' dinosaur than aardvarks, lemurs or kangaroos are typical mammals. It was an animal that had features honed by evolutionary pressures into a form quite unlike that of most other theropods, and even, when taken to an extreme, rather unlike that of most other tyrannosaurs. While its nearest relatives in the *Tarbosaurus* and *Zhuchengtyrannus* genera were very similar, *Tyrannosaurus* stands out as an animal that has undergone a disproportionate amount of research over the decades, and because we consequently know more about *Tyrannosaurus* than any other dinosaur, it has become the best model for future research. Just as the fruit fly *Drosophila melanogaster* is the central base for genetics research, the frog *Xenopus laevis* for neurology and the small roundworm *Caenorhabditis elegans* for developmental biology, *Tyrannosaurus* is a fundamental base for much dinosaur research.[1] While this has obviously contributed to its overexposure in the public eye (and in some scientific circles as well), it also means that it is probably the most studied of any dinosaur. Quite simply, we know more about *Tyrannosaurus* than any other extinct dinosaur, and as a result its biology is a superb topic for discussion (and as a bonus for me, it's therefore an ideal subject to write a book about).

The flipside of this is that I have been forced to refer to *Tyrannosaurus* rather more often than I would have liked in places, simply because it is frequently the only member of the clade for which a given feature or behaviour has actually been confirmed. Not enough research is spread out among other taxa, and while admittedly some are very new (such as *Yutyrannus* and *Lythronax*), and others are known from very little material (*Proceratosaurus*, *Aviatyrannis*), or both (*Nanuqsaurus*), much more work is needed on the anatomy, evolution, and especially the ecology and behaviour of many non-tyrannosaurine tyrannosaurs. Perhaps in part because they are relatively unspecialised, early forms can be somewhat 'lumped in' with animals like smaller megalosaurs or allosaurs when it comes to assessments of probable prey, feeding styles and so on. However, *Tyrannosaurus* is so interesting not just because of what it was, but because of how it got to be there, and the evolutionary trajectories that modified early tyrannosaurs into the incredible animals like the albertosaurines and tyrannosaurines.

The other problem is that dinosaurs in general, and *Tyrannosaurus* in particular, can attract some very odd ideas. No academic field is devoid of the occasional oddball ideas, which may even come from talented and respected researchers, as well as rather more 'fringe' sources. Even if certain controversial issues are finally settled in academic circles, this doesn't necessarily result in them being transmitted more widely; 'Scientists reach agreement' is not half as sexy a story as 'New row over *Tyrannosaurus*'. The public therefore often get to hear only the start of a story, and far too little attention is paid to later work. This, above all else, is why the interminable 'predator versus scavenger' story rumbles on, when it should arguably never have started in the first place and has been crushed in the scientific literature more than once (most thoroughly by the paleontologist Tom Holtz in 2008).[2]

Some of these issues have already been alluded to, while others have been largely overlooked for clarity in the relevant chapters, but they are worth returning to as they are rather commonly misunderstood, or have significant bearing on our understanding of these animals. I'll add here that recent years have seen the media take seriously some ideas that one might charitably call 'intriguing', for example that all dinosaurs lived in water, or that they evolved on other planets in parallel to Earth and are alive and well, having missed the mass extinction on their cosmic home. I don't go into such fringe ideas here (the Internet has them more than covered), but there has been serious discussion in the scientific literature about some plausible theories, and they are hard to overlook. First and foremost is the problem of '*Nanotyrannus*'.

## A tiny tyrant?

In the collections of the Cleveland Museum of Natural History sits a moderately sized theropod skull. It is obviously the skull of a tyrannosaurine: the back of the skull is broad, then tapers rapidly to leave a long but still wide snout with a well-rounded tip, and it has relatively few, large teeth in the jaws. It looks, in fact, quite a lot like a skull of a *Tyrannosaurus*, only it is less than half the size you might expect, being not much more than 50 centimetres long. While it would have belonged to a sizeable animal, it would have been one that was perhaps closer to 5 metres in total length than anything like a typical adult *T. rex*.

Originally described as a specimen of *Gorgosaurus* by palaeontologist Charles Gilmore in 1946,[3] this skull has subsequently been at the centre of much controversy over the years. In part this is because it is a little younger than *Gorgosaurus*, and in fact would have been contemporaneous with *Tyrannosaurus*, but it is also because the skull is not that of a *Gorgosaurus* but of something else. The key question is: did it belong to a juvenile *Tyrannosaurus*, or is it in fact the skull of a miniature tyrannosaurine that lived alongside the most famous of dinosaurs? The latter hypothesis was first formalised in a paper by Bob Bakker and others in 1988, in which they noted that a number of the bones of the skull appeared to be fused.[4]

If that is the case, then this is the skull of an adult individual, and while the animal may have had some growth to do, it was clearly much smaller than any of the other North American tyrannosaurs of the Late Cretaceous, and also deserving of recognition. Benefiting its small size, it was dubbed *Nanotyrannus*.

Since then, debate has raged as to whether or not this is a genuine taxon, as fusion of some skull bones alone is hardly a definitive indicator of adult status. This is important: if the skull represents a new taxon, then *Tyrannosaurus* was not the sole tyrannosaurine of its time in the Americas, and the otherwise large fall-off in size between *T. rex* and the various dromaeosaurs and troodontids is at least partly filled by *Nanotyrannus*, implying a rather different ecology for the carnivores of the time than has been supposed. On the other hand, if this is the skull of a juvenile *Tyrannosaurus*, it gives us an excellent opportunity to examine the growth and development of this genus; with a very young specimen of *Tarbosaurus* already known, there's a great deal of scope for examining how the animals changed over time, and issues of possible ecological separation between juveniles and adults.

Supporters of *Nanotyrannus* as a valid genus point to a few differences in the morphology of the skull that are not seen in specimens of *Tyrannosaurus*. For example, the *Nanotyrannus* skull has several more teeth, but there is always some individual variation in them, and it's not clear how they might have changed as the animals grew. We already know that the limbs changed proportions and that the skull changed shape, so a few other details may well have come and gone during growth. On the other hand, the tooth count did seem to vary in *Gorgosaurus* during growth, and this might apply to *Tyrannosaurus*, too (if admittedly it does not seem to in *Tarboasaurus*) – though tooth count may have been quite variable in tyrannosaurs generally. Moreover, additional analyses, such as those by Thomas Carr,[5] suggest that there are features in common between *Nanotyrannus* and *Tyrannosaurus*, and that the former specimen is a juvenile, not an adult animal.

The issue is further complicated by the presence of 'Jane' (the name, like so many, is honorary, and not a representation of the sex of the individual), a largely complete young tyrannosaurine, which again has been varyingly assigned to *Nanotyrannus* or *Tyrannosaurus* (Fig. 17). Jane was unquestionably a juvenile, having numerous unfused bones throughout the skeleton, as well as some histological evidence showing that it was a young animal, but does it represent another young *Tyrannosaurus* or a second *Nanotyrannus*? Jane was already over 6 metres long, so given that there was apparently some considerable growth still to occur, this can hardly have been a 'dwarf' animal, and furthermore, it had more teeth than a typical adult *Tyrannosaurus*, supporting the idea that they were reduced during growth. Several characters otherwise unique to the skull of *Tyrannosaurus* are also seen in Jane, again supporting the idea that this was just a young *Tyrannosaurus*. However, given the similarity between Jane's skull and the Cleveland skull, this also may suggest that the latter was 'merely' a young *T. rex*.

A final complexity is the problematic specimen recently uncovered in the US, which is in private hands. A small tyrannosaur is preserved alongside a ceratopsian, allegedly representing a fight-to-the-death battle (needless to say, most specialists are highly sceptical), and it has been proposed that this new specimen 'solves' the *Nanotyrannus* question.

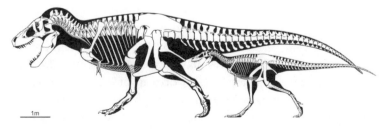

*Fig. 17 Skeleton of the specimen called 'Jane' which by most researchers is considered a juvenile example of* Tyrannosaurus *(adult skeleton shown behind) but has been suggested to represent a small tyrannosaur species. Note the changing skull shape, leg lengths and shape of the pelvis.*

However, although the specimen has been put up for sale it has not been accessible to scientists, so this theory is purely in the realm of speculation at this point. A few imperfect photos of an incompletely prepared specimen are not something to base a judgment on, so for now this remains a frustrating subplot to the overall issue.

Support is growing for both Jane and the Cleveland skull belonging to *T. rex* proper, in part through comparisons with the specimens of very young *Tarbosaurus* from Mongolia,[6] and the growth patterns seen in other dinosaurs. If this assessment is correct, we have an excellent growth series for *Tyrannosaurus*, bolstered further by a small piece of snout housed in Los Angeles, which is from a very small individual, one that was probably only around a year old given the size. In fact, all of this suggests some differences between the tyrannosaurines. Although crushed, the baby *Tarbosaurus* skull looks rather like that of an adult, and suggests that the animal kept a fairly similar skull shape at all ages, simply getting much larger as it grew. Meanwhile, Jane has a skull rather like that of early tyrannosaurs or alioramines (long and thin, and without the broad back of the skull); as it grew, the back 'inflated' to produce the classic skull shape of *Tyrannosaurus*. This points to a major shift in skull function and perhaps ecology as a result. For now, despite some strong objections, it is probably best to consider *Nanotyrannus* invalid, and not a special dwarf tyrannosaur, appealing though the idea may be.

### Two *Tyrannosaurus*?

The *Nanotyrannus* problem is just one of a number of taxonomic issues relevant to *T. rex* being the sole tyrannosaur at the end of the Cretaceous in the Americas, with some specialists advocating that there was a second species of *Tyrannosaurus*. This so-called *Tyrannosaurus* '*x*' was an idea first floated by palaeontologist Dale Russell, though it was Bob Bakker who gave it the '*x*' moniker.[7] It was based primarily on the fact that some specimens of *T. rex* had a pair of small teeth rather than just a single small one at the front of the dentary, and also that some specimens seemed to have rather more robust skulls than others. Based on this and other suggested differences, further researchers have picked up on the idea and proposed that a second *Tyrannosaurus* might be hiding among the *rex* specimens.

At one level this would make sense: it is notable that *Tyrannosaurus* appeared to be the sole large carnivore in its habitat, while both modern mammalian ecosystems and those of dinosaurs typically have two or more large predators, which would leave the *T. rex* system as something of an oddity. However, the data is rather limited, and moreover the differences between the relevant specimens are quite minor. There are certainly differences between the various specimens we have, but we'd expect at least a few to be due to intraspecific variation, and even a few small consistent differences wouldn't necessarily mark out a separate species.

Linked to this issue is the idea that there are two identifiable body types in the known *Tyrannosaurus* material, designated as 'robust' and 'gracile' forms: that is, one is supposed to be more bulky and the other proportionally slender. Moreover, the two body types are not just supposed to be linked to generalised differences in shape, as in people who are stocky or slim; they are said to be connected to subtle sexual dimorphism, with one shape representing males and the other females.[8] As has already been mentioned, a number of dinosaur specimens (but especially those of *Tyrannosaurus*) end up with nicknames, but these are applied pretty much through serendipity and are not meant to relate to the sex of the animal, so 'Sue' is not female any more than 'Bucky' or 'Stan' are male. Previous ideas about distinguishing males from females based on chevron bone numbers or shapes have been shown to be ineffective, and the only reliable way to diagnose a mature female animal is based on the presence of medullary bone. Even here, however, its absence might indicate either that the animal was a male, or that its death did not take place during the breeding season, and not all specimens have been examined.*

So do these 'morphs' even exist, and if they do, do they represent males and females? And if that is the case, which is which? Most researchers remain highly sceptical of these ideas. Data sets are limited, and much of the material is non-overlapping in terms of the parts of the skeleton represented, but it can also be scattered in space and time. Specimens that range across thousands of square kilometres and several million years are all assigned to a species, but they would have represented a number of potentially very different populations. Thus even if there is a signal indicating that specimens can be separated out into two groups, how much is this biased by the problems of such data, and the fact that these animals almost certainly changed in size and shape over the course of their evolution (an individual's growth and variation will also cause issues)?

This is not to rule out either hypothesis, but given the inevitable limitations of such analyses, we'd be looking for much more clear and consistent differences between the two suggested groups. We do see subtle differences between closely related species of all kinds, but even so there are typically some consistent and strong anatomical characters that can be used to tell them apart, and this is fundamental to morphological species concepts that are applied to the dinosaurs. Somewhat inevitably, we need to await additional data: more information should produce more clear-cut results, and with enough specimens it might even be possible to run an analysis of a single population to eliminate many of the issues noted above.

The research rolls on, and although controversies continue to arise and be debated, these do at least often generate additional

---

*For some reason, many museum curators get twitchy when you suggest cutting up their skeletons.

research and the firming up of ideas, as well as the creation of ever-better tests and collections of data to support or refute the contentions at hand. As such, controversial ideas can be useful in stimulating new research; the problems arise when they are clung onto long after they have been debunked. The ideas dealt with here are at least credible, and have been advocated and discussed by serious researchers, though truly 'out-there' ideas also have merit. If nothing else, they demonstrate the perpetual fascination with *Tyrannosaurus* and the attention it receives.

# CHAPTER EIGHTEEN

# The Future

A common cliché in palaeontology is for papers to finish with a sentence along the lines of 'future finds may confirm this hypothesis', which almost inevitably means that the paper is frustrating in its lack of firm conclusion (readers will note that I finished the last section on a variation of this very phrase). Some excellent, ingenious studies or logical progressions through the available data and evidence can provide wonderful insights into the biology of extinct animals, but to confirm these – no matter how plausible they might seem – we need hard evidence. As mentioned previously, that kind of hard data or convincing confirmation is often absent, so we are left with unanswered questions and hanging possibilities.

Fortunately, however, as palaeontological research progresses ever more specimens are recovered in the field, resulting in increasingly detailed studies on those bones, teeth, skins and feathers, and this information is communicated ever more effectively between researchers. The idea that 'this is a golden age of science' will perhaps always be true, since technology, new ideas and developments continue apace, but clearly science benefits from such changes and palaeontology is very much more advanced today than it was even a decade ago, let alone 30–40 years in the past. As a result, we are already confirming long-held hypotheses in tyrannosaur palaeontology, and future discoveries are likely to confirm still further ideas.

For example, in the original notes for this chapter I wrote that we may yet find true dwarf tyrannosaurids, most probably in an island setting – but before I got around to writing this section fully, *Nanuqsaurus* had been discovered. My original theory was not the result of any brilliant prognostication on my part,[*] but arose naturally from the application of

---

[*]Though feel free to credit me in this regard.

fundamental aspects of biology. Dwarf forms of large-bodied lineages often appear on islands, we already know of dwarf dinosaurs that existed on islands, and the large tyrannosaurs were around for a long time, including in times and places where there were islands. This therefore makes it likely that a dwarf tyrannosaurid evolved at some point – but actually finding a fossil of one is a rather more difficult proposition. Populations on islands would have been small (there's not that much room on them), islands are not usually great places for large amounts of deposition to take place (though the sea around them may be better for this), and island populations are vulnerable to going extinct. These factors combined together make it improbable that such discoveries would be made, even if several dwarfs were around, so it was rather a surprise for me to be gazumped at short notice by the little tyrannosaur from the north. On the other hand, this event does demonstrate the point I am making. A solid prediction based on a grounded understanding of biological and evolutionary patterns can go a long way, and can potentially be confirmed despite the limits of the fossil record.

This chapter, then, looks at some possible future developments in the fields of tyrannosaur research. At various points in this book I have introduced my own ideas on aspects of tyrannosaur biology that are not in the scientific literature as formal hypotheses, but that are plausible or even likely given what we know about these animals and biology in general. Here, though, are a few more theories that do not fit into the specifics of tyrannosaur evolution, but that represent some plausible ideas about evolutionary lineages and discoveries that may yet appear.

## Things to come

For a start, there are some basics that are perhaps inevitable. We do now have, for example, huge numbers of fossil eggs for all manner of dinosaurs, including a decent number of theropods, and despite the overall rarity of tyrannosaur fossils we can expect to find a nest that can be confidently attributed to a tyrannosaur. With luck there might even be a brooding adult, or at least a guarding one, alongside, and perhaps even an embryo or two inside the shells. A find such as this would help in diagnosing other tyrannosaur eggs (some probably already exist, but telling which dinosaur eggs belong to which group without something like an embryo to start the process is all but impossible). It would also add a really fundamental point to the current data on growth and reproduction. The nest would tell us much about how many eggs were laid, how large they and the hatchlings were, how quickly the latter would have grown, and what developmental changes might have taken place early on in the life of a baby tyrannosaur, adding to our store of knowledge about the animals' taxonomic characters, and their behaviour and ecology.

Another inevitability is that we will find many new species. As has already been mentioned, at least one species of tyrannosaur that is considered new has yet to be named, and others will follow. Areas like southern China, Mexico, eastern Russia and eastern Canada should all support tyrannosaur fossils, but they either have relatively few fossils beds, or have yet to be extensively excavated and studied. It is therefore likely that a number of new finds will flow from these locations in the not too distant future. Canada and Mexico are especially intriguing. Might *Tyrannosaurus* or its immediate predecessors have had much wider ranges than we currently know of? Were there distinct species, or perhaps even distinct lineages, in other North American areas? Teeth assigned to *Tyrannosaurus* are known from northern Mexico, and although so far only isolated bones have been found, the possibility that they belonged to a new species or a different genus cannot be ruled out.

Similarly, there are areas that should yet reveal the presence of tyrannosaurs where we might expect them to be present. *Stokesosaurus* is known from the extraordinarily productive Late Jurassic Morrison Formation beds of the US, but in Tanzania, in East Africa, there is a parallel set of fossils from the area known as Tendaguru. When the material from the African location was first uncovered, many of the species identified were assigned to the same genera as those from the

American beds, including *Brachiosaurus* and *Dryosaurus*. A number of these have since been reclassified as their own genera, but this example does demonstrate how similar in general the two faunas were, and as a result, something tyrannosaurian should be expected. A recent study of numerous isolated theropod teeth suggested that all manner of different theropod clades were present at Tendaguru, yet none of those identified so far were obviously tyrannosaurian in origin. Perhaps the tyrannosaurs were simply absent, but if they were already in Asia, Europe and North America, and there were apparent close similarities between various faunas, then it is a little surprising that they have yet to appear here.

On a related note, we do know of a number of tyrannosaurs from very fragmentary fossil remains. While additional material for animals like *Zhuchengtyrannus* and *Nanuqsaurus* would be welcome, at least their fossils are diagnosable as distinct taxa. The only known specimen of *Raptorex* is a juvenile and it would be great to have an adult animal, and something more than just a head for *Proceratosaurus* would be nice. The partial pubis from Australia has had repeated mentions in this book, and this is because it is both intriguing and important, however, we would also very much like to see the rest of it. More material would help to determine if it really did belong to a tyrannosaur (personally, I think it did), which lineage it was from, and what implications this would have for both the diversity and the dispersal of the tyrannosaurs. If the tyrannosaurs did spread that far and took up residence in the southern continents relatively early on in their evolutionary history, then we can expect a lot more taxa to follow at some point; it is unlikely that the tyrants would have made it to Australia and never radiated. Currently, dinosaur fossils from Antarctica itself are extremely rare because of the difficult conditions there, but if tyrannosaurs were plentiful down south, then it is likely that they occupied this region at some point as well, and again, we might expect to find fossils of new species associated with their spread.

Moving on to interactions between animals, we do already have several dinosaur fossils with evidence of damage from tyrannosaurs, and plenty of tyrannosaurs show injuries inflicted by other tyrannosaurs, but to date there's no good indication of injuries to them inflicted by their intended meals. Naturally, predators do go out of their way to avoid taking on dangerous opponents, but young and inexperienced carnivores do make mistakes themselves, and animals can always get unlucky. However, although such injuries would have been rare, they would have been dealt out at least occasionally. A find demonstrating such an occurrence would certainly be spectacular, if only because it is so unlikely to ever turn up, but somewhere out there a fossil *Tyrannosaurus* with a *Triceratops* horn stuck through it may exist. Rather more likely would be finding some common pattern in injuries, such as blows to the lower legs showing evidence of impact, which might lead to the conclusion that ankylosaurs defended themselves against giant, bone-crushing carnivores (and at least one *Gorgosaurus* had a very nasty injury on the shin).

More information from stomach contents and/or bite marks will also almost inevitably come with additional specimens of both tyrannosaurs and other animals. Again, young animals would be especially welcome here; it would be great to see if younger tyrannosaurs really did take different prey from that of their full-sized parents, and in particular if they took non-dinosaurian prey. It's most likely that they generally ate lizards, frogs, small mammals, and maybe even fish and insects when very small, and stomach-content data would probably confirm that they did eat these things, even if it would not make it clear if this was a genuine preference. Very young tyrannosaurs may even have consumed bits of bones from large dinosaurs they could not possibly have killed or broken up: fragments of very large bones in their stomach contents would be good indicators that a parent was smashing up food for its offspring to eat.

## Animals of a feather

The discovery of more feathers for tyrannosaurs is also a near certainty. Until recently, feathers (or impressions of them) were pretty much only preserved in exceptional circumstances, in beds that contained super-fine ash or similar sediments that could trap and preserve these things. Such preservational conditions do also tend to flatten specimens, so there is largely a trade-off between getting a flat specimen with feathers (and beaks, claws, hair, skin and other soft tissues for your animal of choice), or a 3D specimen of just bones and teeth. There have been odd exceptions with incredible preservation of soft tissues and non-flattened bones in such beds, but not typically the reverse. However, recently a young ornithomimosaur has been found in Canada which was preserved in more 'normal' sandstone beds, and has both good bone preservation and feathers. This is remarkable, as the fossil beds in question were not thought capable of preserving such soft tissues. More importantly in the context of this book, they include beds that house *Tyrannosaurus*. It is looking ever more likely that we will find the king with his plumes one day. The increasing number of feathered specimens from China also points to the likelihood of getting additional material for *Dilong* or *Yutyrannus*, as well as perhaps feathered juveniles.

Moving on to more major evolutionary changes, *Nanuqsaurus* may merely be the first dwarf genus to have turned up. It is not impossible that others lived alongside it in similar isolation, and we do know of island systems that Late Cretaceous dinosaurs occupied, producing dwarf forms of various lineages (most notably in modern Romania), so presumably there are others still waiting to be discovered. *Nanuqsaurus* is also interesting because it appears to buck another classic evolutionary pattern: Bergmann's Rule of increasing size in high latitudes, described earlier. *Tyrannosaurus* and at least some of the albertosaurines inhabited relatively northern latitudes, and encountered snow and cold conditions, so there is no immediate reason to think that they could not have roamed much further north; thus there may yet be still larger taxa to find, or perhaps larger populations of these already huge animals (and, conversely, maybe those down in Mexico were a little smaller).

There is something of an overemphasis on the size of the tyrannosaurs in the popular literature, and while the size that they achieved was incredible, there is much more to them than just how big they were. Large size in an animal is interesting, not so much because of its impressiveness, but because size determines so much of an animal's biology, and the often interesting compromises and issues that come with being big. Nevertheless, it's nice to realise that however huge tyrannosaurs were – and Sue was certainly massive – considerably larger individuals are likely to have existed. We have decent skeletons from a limited number of adult individuals, and scattered remains of a few dozen more. What are the chances that in this selection we have evidence of, say, the top 1 per cent of animals by size, let alone the largest individuals ever? It is again impossible to know without direct evidence, but I would be very surprised if individuals did not exist that weighed well over 10 tonnes and were more than 15 metres in total length, and quite possibly a fair bit more.

At the other end of the size spectrum there should also be more small taxa to find. Suffering the joint bias of being both old and small, fossils of early tyrannosaurs from the Middle and Late Jurassic (and to a lesser degree the Early Cretaceous) are likely to remain elusive for some time, but we can expect more of them to appear. It is extremely unlikely that the limited number of Middle Jurassic tyrannosaurs we know of today – already spread across thousands of kilometres of continental land mass – represented all there was of this radiation. The origins of groups often coincide with their exploitation of a new niche, or takeover from some previous group, and we might therefore expect that these animals did spread and diversify rapidly, in which case there may be many more species to discover.

There may even be entire new lineages to find; the proceratosaurids were only recognised as a group very recently, and some recent pylogenies suggest that the alioramines represent a separate branch of tyrannosaurid evolution outside of the normal albertosaurine-tyrannosaurine split. Such rearrangement of the tyrannosaur tree, or additions to it, would be especially interesting, because this would provide new opportunities to appreciate evolutionary changes, and by extension evolutionary pressures that led to these changes. Seeing how groups changed and working out what may have prompted the changes – for example the origins of feathers, large size, movement between continents, and new jaw and tooth shapes – are key areas of biology and ecology, and investigation in these areas provides some of the most fascinating aspects of paleontology.

Moving on, there are a number of studies that are ongoing or increasing in prominence in palaeontology that will have a potentially major effect on our understanding of tyrannosaurs. Obviously, there will be ever better and more detailed descriptions of the anatomy of these animals, providing new information and data, and this will be plugged into more comprehensive and exacting phylogenetic analyses, allowing us to track character changes and evolutionary rates with greater precision. Ongoing work on growth rates and ontogenetic changes will improve our understanding of the animals' physiology and metabolic processes.

### Any colour you like?

There has recently been some incredible and highly controversial work on exceptionally well-preserved tissues of non-avian dinosaurs, including *Tyrannosaurus*, from Montana, in the US.[9] So far only fragments and bits of material have been described, but the researchers are confident that real soft tissues have been preserved, despite the colossal lengths of time involved. Scepticism remains over how such preservation might be possible, but recent published work describes a feasible mechanism for the long-term survival of the material, adding credibility to the idea that these *are* genuine bits of soft tissue that have been largely unaltered for tens of millions of years. There is nothing yet published on individual cells, let alone the possibility of genetic material, but given the amount of specimens apparently being worked on by the research group, we can at least expect some surprises and groundbreaking finds on biological material for dinosaurs.

Exceptional preservation of other material also provides potential new data streams for tyrannosaurs. Although we can expect to get new specimens of feathered tyrannosaurs, those that already exist can still yield new data. Work on feathered dinosaurs and birds from the Jurassic and Cretaceous of eastern China (and other studies of similarly preserved feathers from Brazil and Germany) examines the microstructure and chemical signatures of these feathers. Chemical traces can suggest certain pigment types, and by extension possible colours and patterns, while the colours of many feathers are also linked in part to the shapes of microstructres called melanosomes that are preserved in some feathers. Successes with techniques to assess these traces give insights into the plumage of long-extinct animals.

There are no absolutes here; we are talking in general terms (such as reds versus browns versus blacks, rather than specific tones, and some colours cannot be easily determined), but it is still remarkable that we are taking the first steps towards understanding dinosaur colours, and having some empirical evidence for them. There is no specific reason to think that these methods could not be applied to animals like *Yutyannus*, and it is perhaps only a matter of time before we see papers describing the possible colours and patterns of at least one tyrannosaur, and all of the implications that will have. Might the animals have had bold patterns and colours for signalling, or muted tones and patterns for use in camouflage? Did juveniles have different patterns from adults, with different emphasis on different aspects of their lives (they might have been more concerned with avoiding being eaten themselves, and would certainly have had no need to find and woo a mate)? Might there have been different-coloured males and females? *Yutyrannus* is thought to have lived in a pretty cold environment, so maybe it shed its coat and switched to a different shade and pattern in winter, like so many northern animals do today. The possibilities are both intriguing and tantalising, given that the data almost certainly already exists to begin at least some of the analyses to test these ideas.

Technological changes are at the forefront of scientific developments, and though much of palaeontology still relies on the methods applicable 100 or even 200 years ago (books of anatomy in the lab, a hammer in the field), CT scans, 3D reconstructions, structural analyses of bone deformation and other such analyses are commonplace, and allow the collection and analysis of data considered impossible little more than a decade ago. Studies of such mechanically complex areas as standing, walking and biting are increasingly being carried out, as both the ability to scan fossils and the computing power to analyse the effects on them and interactions between them increase. With further increases in analytical ability, more dynamic analyses will be possible. For example, there are already a number of studies examining bite forces, how bones and teeth react under loads, how such things affect the structure of the skull, and so on, but the analyses are being applied to a static subject: something like a hypothetical single and unmoving bone. Analysis of a dynamic bite, where the tyrannosaur and prey item might both move in complex ways, would give far more nuance and have the potential to be matched up to known bite marks and feeding traces. Did tyrannosaurs bend and twist the neck when biting, and could they resist the struggles of a baby dinosaur desperate to escape? Certainly, additional integration of such data would be fascinating from the mechanical point of view, and it has the potential to link up anatomical, behavioural and ecological data at once, and link to the under-utilised trace fossils from bites on bones.

Computing power is also increasingly brought to bear in the field of ecology, with complex models analysing changes across entire populations of interacting organisms within ecosystems. Some simple models have been applied to paleontology, including possible tyrannosaur habits, but a lack of data is often crippling; we really don't have many useful ideas, let alone hard data, for birth rates, population density, numbers of adults versus juveniles, and other basic and important information needed for such work. However, as new data and understanding accumulates in the ecology of living animals, it will become easier to at least place limits on possibilities for tyrannosaurs, backed by new fossil data. Information on feeding interactions (including what tyrannosaurs ate, and also which herbivores ate which plants) will help to restore possible food webs, data on growth and physiology will enable evaluations of resource requirements, discoveries about nests will assist with assessments of possible population structures, and so on. Little will be concrete, and the margins of error are likely to be huge, but the first steps towards reconstructing a real dinosaurian ecosystem as a functioning and biological unit, rather than just a list of species and who might have eaten what, is becoming increasingly likely. Tyrannosaurs are likely to be a key part of this since some of the best-studied and understood ecosystems, with the most data, include them, they will be a fundamental part of such work, and we do broadly have a good idea of how interactions can operate in large terrestrial organisms.

In short, on top of the inevitable procession of new data and analyses, and ever improving techniques, there are some strong candidates for major advances in several fields in the coming years. Due to the charisma of the tyrannosaurs they may perhaps receive a disproportionate amount of attention, but we can expect these animals in particular to benefit from such research, and the rate at which our understanding increases, and the depth and breadth of study, are only set to increase. Areas considered permanently out of bounds even a few years ago, like identifying colour in feathers, are starting to become not only a possibility but an inevitability (if with limits due to a hundred-odd million years of ravages to the data), and the only real certainty is that there will be developments and discoveries far beyond what I'm willing to speculate about here. I can't wait.

# CHAPTER NINETEEN

# Conclusions

The end of the tyrannosaurs came, as for so many dinosaurs, at the end of the Cretaceous Period. The last of the tyrannosaurs that we know of includes *Tyrannosaurus*, this genus appeared right up to the end of the Cretaceous and was one of the very last non-avian dinosaurs that we know of. Certainly, some other tyrannosaur species must have been around somewhere, but we don't have too many fossil beds that close to the boundary of the end of the Cretaceous, so once more we are a little short of data. Given the presence of so many tyrants in Asia in the later parts of the Cretaceous, it's likely that there were a number of species still hanging around the Gobi and perhaps in Europe and the southern parts of North America.

Around 66 million years ago, it is clear that there was a massive change in the make-up of life on Earth: all the non-avian dinosaurs became extinct: there were no more tyrannosaurs, no sauropods, ceratopsians, hadrosaurs, ankylosaurs or any others. Also failing to make it were a great many of the birds (just a few lineages made it), various mammals, many crocodilians, champsosaurs, pterosaurs, the marine plesiosaurs, and huge numbers of other animals and plants. There was, in fact, a mass extinction that affected species worldwide. In geological terms this was close to something instantaneous, lasting perhaps just a few tens of thousands of years.

It is this change that marked the end of the Cretaceous and indeed the Mesozoic Era as a whole. Early geologists used the changing faunas to help separate out the various periods of time, and clearly this was a sudden and dramatic change. More recent work has naturally refined that division, and it is the rocks themselves and their differences that now mark the shift and the end of the Mesozoic. Many hypotheses have been advocated for the end of the dinosaurs, most of which can only be described as fringe, or at least the product of rather loose thinking. A variety of popular theories have been put forward over the years: the dinosaurs became too large to support themselves, they became too senile as a race and forgot to breed, mammals ate all their eggs, they all died out suddenly due to some global pandemic – there are plenty more odd and certainly interesting ideas. Most can be discounted as quite silly, and some might (just, possibly) explain the end of the dinosaurs, but not the loss of so many other varied and widespread lineages. Only two have really stood the test of time and can now be considered viable hypotheses for the end of the dinosaurs: mass vulcanism and meteoric impact.

Both of these clearly occurred. There were mass volcanic eruptions in India at the end of the Cretaceous that saw huge volumes of lava, ash and gasses ejected into the atmosphere, and created the Deccan Traps ash beds in India. Equally certainly, a massive meteor did hit the Earth, just off the north-east coast of modern Mexico, at the very end of the Cretaceous, and would similarly have thrown up huge volumes of dust and other matter. Both these events would have caused major climate changes: atmospheric particles would have blocked the sun, chilling the Earth and killing off both plants and animals, which due to the rapidity of the changes were unable to adapt or get enough food to survive. With the loss of plants especially, the herbivores would not have survived, and as a result the carnivores, and whole food chains and ecosystems, would have collapsed.

The question is whether many systems, and especially dinosaur-dominated ones, were already on their last legs due to the actions of the volcanoes before the meteor finished them off, or whether they managed to tolerate these changes only to be blindsided by the impact of the meteor. Currently, the latter scenario seems the most plausible, and although the eruptions would certainly have had massive local effects in Asia, they don't seem to have been a global event in the way that the meteor impact was.

The initial local event of the meteor crash would have been phenomenal, generating earthquakes, tsunamis, falling debris and perhaps enough heat to set fire to the air itself. Little would have survived for quite some distance around the impact, and effects like tidal waves could have had serious local effects across the oceans. Enough dust was thrown into the atmosphere that even in Italy, many thousands of kilometres away, there is a red layer of iridium-rich clay visible in the rocks. Iridium is rare on Earth but common in meteors and asteroids; that found in Italy comprises the remnants of the detonation in Mexico, still visible halfway around the world. It's easy to imagine just how devastating this event was.

Life survived, of course: despite the colossal losses, numerous animals and plants made it. They probably survived in low numbers, with some few individuals or populations living in sheltered places being lucky enough to miss the worst of things, and others perhaps being well adapted already to harsh conditions, or evolving quickly enough to survive. Mass extinctions always contain puzzles: some things that should die off survive, and some that you would expect to survive are lost – but predictions can be made in this context. For a start, survivors tend to be small. This means that they tend to have large populations (with both more individuals and more genetic diversity), reproduce rapidly (so can recover from a population crash quickly) and require few resources to keep going. Those that can migrate quickly and thus reach other areas will probably do relatively well, and those that are generalists, already living in a variety of ecosystems and capable of taking a variety of foods, are likely to fare less badly than more specialised animals.

Thus it is relatively easy to see why many dinosaurs may have been doomed. They did have large numbers of offspring and had a global distribution collectively, but many were large, and it would have been years before a new generation could have bred again following a population crash. Many were relative specialists, and of course in the case of the tyrannosaurs, were also carnivores. They had little chance in the grand scheme of things. Nevertheless, it's unlikely that they all became extinct instantly. From time to time there are reports of dinosaur bones beyond the Cretaceous, and although none has yet turned out to be correctly attributed (turnover of soil by roots of plants and similar phenomena seem to be responsible for such errors), it really would not be a surprise if a small number of dinosaurs did limp on for a few thousand years. Somewhere there may well have been a valley or island that was largely free of the devastation, and perhaps some dinosaur species did keep on going there for a while, but a small population in a small area would always be vulnerable to some small-scale disaster and dying out. I would not be at all surprised if one day a genuine non-avian dinosaur find was made in the rocks that come after the end of the Cretaceous, but it would cast no doubt on the extinction event at the end of the Cretaceous, or the overall devastation and death of the dinosaurs. One lone bone (or even evidence of a few species) would not make the science questionable or the dinosaurs any less dead, or give credence to a sauropod living in the Congo (or, for that matter, pterosaurs existing in New Guinea, or the Loch Ness Monster living in Scotland, despite some suggestions to this effect).

The end of the dinosaurs allowed the eventual flourishing of the lineages that survived the transition out of the Mesozoic. In a geological sense this was literally the dawning of a new era, but the tyrannosaurs were gone for good, leaving behind only their fossil remains buried deep in the Earth.

The story of the tyrannosaurs covers more than a century of research, with dozens of palaeontologists contributing hundreds of papers describing around 30 species of tyrannosaur, which are represented by hundreds of fossils from dozens of localities on five continents. Over the course of this book we have covered aspects of the origin, evolution and diversity of tyrannosaurs, their ecology and behaviour, their anatomy and functional morphology, and their extinction. Even so, it is barely possible to even mention all of the hypotheses that have been advocated for members of this group at one time or another, let alone deal with the often long and detailed cases made for or against a given idea.

The job of science is to sort out the hyperbole from reality, or at least to try to move away from raw hypotheses into testing them. From there we can sort out the possible from the impossible. Huge amounts of time and effort are invested in testing the ideas, sometimes involving ingenious solutions to the problems of conducting tests on fossil bones and teeth that are a hundred million years old. Recent years have brought fourth new technologies to apply to the analyses, such as CT scans of skulls to analyse their brains, computer modelling of standing, turning and running animals, and stress models of skulls. Detailed studies of living species and how they evolve and behave provide models for comparisons to extinct carnivores, and naturally all this is backed by constant new fossil finds and interesting discoveries.

Collectively, this work is giving us a much clearer picture of tyrannosaurs than we have ever had before. Ideas like dedicated scavenging, and larger females than males, have been assessed and rejected; others, such as cranio-facial biting and consumption of juvenile dinosaurs, have found support. Compared with even a decade ago, nearly twice as many tyrannosaur species are known, analyses of their relationships are many times larger and more detailed, and we now have direct evidence of both scavenging and predation, and of cannibalism. We have specimens of very young tyrannosaurs that reveal ideas about growth, preservation of feathers, specimens from South America and Australia, very early tyrannosaurs, data on head crests and their possible uses in signalling, coprolites and stomach contents, and, most dramatically of all, tantalising hints about the preservation of soft tissues like ligaments and muscles.

Tyrannosaurs were much more than near-armless bone crushers, but they were also real animals. We need to bear in mind that for all that seems odd and amazing about them, like the sheer size of the tyrannosaurines, they fitted into their world just as much as any other organism has ever done, from a bacterium to the smallest flower and largest whales. They had their ancestors and descendants, competitors, predators, diseases, parasites and prey. Their biology and behaviour influenced that of numerous other species around them, and helped drive changes to both flora and fauna for millions of years. In turn, they were influenced by changing environments and species that they interacted with. In that sense at least, they were not special, but 'merely' another of an apparently endless succession of forms that have occupied this planet and continue to do so. From this perspective it is easier to appreciate how different they truly became compared with even their near relatives in the Theropoda.

The rise of the tyrannosaurs stemmed from some relatively unremarkable animals of the Middle Jurassic that went on to produce some of the largest, heaviest and most anatomically unusual carnivores of all time. Chronicling the tyrannosaurs and putting them in the context of their geological time and evolutionary place provides a wonderful example of modern palaeontology. It involves data drawn from a wide variety of sources, and analyses that encompass not just biology and geology – the fundamental fields of the discipline – but also chemistry, physics and engineering. Discoveries from around the world are pieced together by palaeontologists, and fragments of data can be used to fill in gaps in ideas, and start to complete the jigsaw of our understanding and knowledge of this group.

I am trying at this point not to end on the cliché phrase 'more data is needed', and much as it is tempting to do so, I'll try instead to finish on this note. The last 15–20 years have seen perhaps the biggest upheaval and advancement of palaeontological data, and techniques that are more advanced than those of the previous 150–200 years, and the rate of increase in knowledge is increasing ever more quickly. It is not just that we now know more about these animals than at any point in the past (another overused phrase), but that we are approaching something of a critical mass, where each new piece of data or study fits into the existing framework, rather than being left hanging in intellectual space, unclear of its relationship to other concepts. In short, we don't just know more, and we are not learning more quickly; it's that we now have a good handle on what we know and what we don't, and on how to go about filling in the gaps in our knowledge. It is not, then, that more data is needed – it is that more data is coming, which will help answer questions and further expand on these tyrannosaur chronicles.

# References

## Introduction

1. Osborn, H. F. 1905. *Tyrannosaurus* and other Cretaceous carnivorous dinosaurs. *Bulletin of the American Museum of Natural History*, 21:259–265.
2. Milner, A. R., Harris, J. D., Lockley, M. G., Kirkland, J. I. & Matthews, N. A. 2009. Bird-like anatomy, posture, and behaviour revealed by an Early Jurassic theropod dinosaur resting trace. *PLOS ONE*, 4:e4591.
3. Hendrickx, C., Hartman, S. A. & Matteus, O. 2015. An overview of non-avian theropod discoveries and classification. *PalArch*, 12:1–73.
4. Holtz, T. R. 2004. Tyrannosauroidea. In: *The Dinosauria*, 2nd edn, Weishampel, B., Dodson, D., Osmólska, H. (eds), University of California Press, Berkeley, pp. 111–136.
5. Rauhut, O. W. M., Milner, A. C. & Morre-Fay, S. 2010. Cranial osteology and phylogenetic position of the theropod dinosaur *Proceratosaurus bradleyi* (Woodward, 1910) from the Middle Jurassic of England. *Zoological Journal of the Linnean Society*, 158:155–195.
6. Holtz, T. R. 2004. Tyrannosauroidea.
7. Ibid.
8. Lü, J., Yi, L., Brusatte, S.L., Yang, L., Li, H. & Chen, L. 2014. A new clade of Asian Late Cretaceous long-snouted tyrannosaurids. *Nature Communications*, 5:3788.
9. Currie, P. J. 2003. Cranial anatomy of tyrannosaurid dinosaurs from the Late Cretaceous of Alberta, Canada. *Acta Palaeontologica Polonica*, 48:191–226.
10. Brusatte, S. L., Hone, D. W. E. & Xu, X. 2013. Phylogenetic revision of *Chingkankousaurus fragilis*, a forgotten tyrannosaur specimen from the Late Cretaceous of China. In: *Tyrannosaurid Paleobiology*, Parrish, J. M., Molnar, R. E., Currie P. J., Koppelhus, E. B. (eds), Indiana University Press, Bloomington, pp. 2–13.

11. Hone, D. W. E., Wang, K., Sullivan, C., Zhao, X., Chen, S., Li, D., Ji, S., Ji, Q. & Xu, X. 2011. A new, large tyrannosaurine theropod from the Upper Cretaceous of China. *Cretaceous Research*, 32:495–503.

12. Hone, D. W. E. & Watabe, M. 2010. New information on the feeding behaviour of tyrannosaurs. *Acta Palaeontologica Polonica*, 55:627–634.

13. Novas, F. E., Agnolín, F. L., Ezcurra, M. D., Canale, J. I. & Porfiri, J. D. 2012. Megaraptorans as members of an unexpected evolutionary radiation of tyrant-reptiles in Gondwana. *Ameghiniana*, 49 (suppl.):R33.

14. Loewen, M. A., Irmis, R. B., Sertich, J. J., Currie, P. J. & Sampson, S. D. 2013. Tyrant dinosaur evolution tracks the rise and fall of Late Cretaceous oceans. *PLOS ONE*, 8:e79420.

15. Xu, X., Clark, J. M., Forster, C. A., Norell, M. A., Erickson, G. M., Eberth, A., Jia, C. & Zhao, Q. 2006. A basal tyrannosauroid dinosaur from the Late Jurassic of China. *Nature*, 439:715–718.

16. Benson, R. B., Barrett, P. M., Rich, T. H. & Vickers-Rich, P. 2010. A southern tyrant reptile. *Science*, 327:1613.

17. Lloyd, G. T., Davis, K. E., Pisani, D., Tarver, J. E., Ruta, M., Sakamoto, M., Hone, D. W. E., Jennings, R. & Benton, M. J. 2008. Dinosaurs and the Cretaceous terrestrial revolution. *Proceedings of the Royal Society of London B: Biological Sciences*, 275:2483–2490.

## Morphology

1. Currie, P. J. 2003. Cranial anatomy of tyrannosaurid dinosaurs from the Late Cretaceous of Alberta, Canada. *Acta Palaeontologica Polonica*, 48:191–226.

2. Rayfield, E. J. 2004. Cranial mechanics and feeding in *Tyrannosaurus rex*. *Proceedings of the Royal Society of London B*, 271:1451–1459.

3. Stevens, K. A. 2006. Binocular vision in theropod dinosaurs. *Journal of Vertebrate Paleontology*, 26:321–330.

4.  Brochu, C. A. 2000. A digitally rendered endocast for *Tyrannosaurus rex*. *Journal of Vertebrate Paleontology*, 20:1–6.

5.  Abler, W. L. 1992. The serrated teeth of tyrannosaurid dinosaurs, and biting structures in other animals. *Paleobiology*, 1992:161–183.

6.  Snively, E., Russell, A. P., Powell, G. L., Theodor, J. M. & Ryan, M. J. 2014. The role of the neck in the feeding behaviour of the Tyrannosauridae: inference based on kinematics and muscle function of extant avians. *Journal of Zoology*, 292:290–303.

7.  Currie, P. J. 2003b. Allometric growth in tyrannosaurids (Dinosauria: Theropoda) from the Upper Cretaceous of North America and Asia. *Canadian Journal of Earth Sciences*, 40:651–665.

8.  Persons, W. S. & Currie, P. J. 2011. The tail of *Tyrannosaurus*: reassessing the size and locomotive importance of the *M. caudofemoralis* in non avian theropods. *The Anatomical Record*, 294:119–131.

9.  Schachner, E. R., Hutchinson, J. R. & Farmer, C. G. 2013. Pulmonary anatomy in the Nile crocodile and the evolution of unidirectional airflow in Archosauria. *PeerJ*, 1:e60.

10. Dal Sasso, C. & Maganuco, S. 2011. *Scipionyx samniticus* (Theropoda: Compsognathidae) from the lower Cretaceous of Italy. *Memorie della Società Italiana di Scienze Naturali e del Museo Civico di Storia Naturale di Milano*, 37:1–283.

11. Nesbitt, S. J., Turner, A. H., Spaulding, M., Conrad, J. L. & Norell, M. A. 2009. The theropod furcula. *Journal of Morphology*, 270:856–879.

12. Xu, X., Clark, J. M., Mo, J., Choiniere, J., Forster, C. A., Erickson, G. M., Hone, D. W. E., Sullivan, C., Eberth, D. A., Nesbitt, S., Zhao, Q., Hernandez, R., Jia, C., Han, F. & Guo, Y. 2009. A Jurassic ceratosaur from China helps clarify avian digital homologies. *Nature*, 459:940–944.

13. You, Y. Sullivan, C. & Xu, X. 2015. Three-dimensional modelling of the manual digits of the theropod dinosaur

*Guanlong*, with a preliminary functional analysis. *Acta Palaeontologica Sinica*, 54:165–173.

14. Snively, E. & Russell, A. P. 2003. Kinematic model of tyrannosaurid (Dinosauria: Theropoda) arctometatarsus function. *Journal of Morphology*, 255:215–227.

15. Manning, P. L., Ott, C. & Falkingham, P. L. 2008. A probable tyrannosaurid track from the Hell Creek Formation (Upper Cretaceous), Montana, United States. *Palaios*, 23:645–647.

16. Hutchinson, J. R. & Allen, V. 2009. The evolutionary continuum of limb function from early theropods to birds. *Naturwissenschaften*, 96:423–448.

17. Xu, X., Wang, K., Zhang, K., Ma, Q., Xing, L., Sullivan, C., Hu, D., Cheng, S. & Wang, S. 2012. A gigantic feathered dinosaur from the Lower Cretaceous of China. *Nature*, 484:92–95.

18. Birn-Jeffery, A. V., Miller, C., Naish, D., Rayfield, E. J. & Hone, D. W. E. 2012. Pedal claw curvature in birds, lizards and Mesozoic dinosaurs – complicated categories and compensating for mass-specific and phylogenetic control. *PLOS ONE*, 7:e50555.

19. Godefroit, P., Golovneva, L., Shchepetov, S., Garcia, G. & Alekseev, P. 2009. The last polar dinosaurs: high diversity of latest Cretaceous arctic dinosaurs in Russia. *Naturwissenschaften*, 96:495–501.

20. Erickson, G. M., 2014. On dinosaur growth. Annual Review of Earth and Planetary Sciences, 42:675–697.

21. Erickson, G. M., Makovicky, P. J., Currie, P. J., Norell, M. A., Yerby, S. A. & Brochu, C. A. 2004. Gigantism and comparative lifehistory parameters of tyrannosaurid dinosaurs. *Nature*, 430:772–775.

22. Hone, D. W. E. 2012. Variation in the tail length of non-avian dinosaurs. *Journal of Vertebrate Paleontology*, 32:1082–1089.

23. Hutchinson, J. R., Ng-Thow-Hing, V. & Anderson, F. C. 2007. A 3D interactive method for estimating body segmental parameters in animals: application to the turning and running performance of *Tyrannosaurus rex*. *Journal of Theoretical Biology*, 246:660–680.

24. Rauhut, O. W. M., Milner, A. C. & Morre-Fay, S. 2010. Cranial osteology and phylogenetic position of the theropod dinosaur *Proceratosaurus bradleyi* (Woodward, 1910) from the Middle Jurassic of England. *Zoological Journal of the Linnean Society*, 158:155–195.

25. Holtz, T. R. 2004. Tyrannosauroidea. In: *The Dinosauria*, 2nd edn, Weishampel, B., Dodson, D., Osmólska, H. (eds), University of California Press, Berkeley, pp. 111–136.

26. Fiorillo, A. R. & Tykoski, R. S. 2014. A diminutive new tyrannosaur from the top of the world. *PLOS ONE*, 9:e91287.

## Ecology

1. Clark, J. M., Norell, M. A. & Chiappe, L. M. 1999. An oviraptorid skeleton from the Late Cretaceous of Ukhaa Tolgod, Mongolia, preserved in an avian-like brooding position over an oviraptorid nest. *American Museum Novitates*, 3265:1–36.

2. Ibid.

3. Horner, J. R. 2000. Dinosaur reproduction and parenting. *Annual Review of Earth and Planetary Sciences*, 28:19–45.

4. Erickson, G. M., Currie, P. J., Inouye, B. D. & Winn, A. A., 2006. Tyrannosaur life tables: an example of nonavian dinosaur population biology. Science, 313:213-217.

5. Erickson, G. M., Makovicky, P. J., Currie, P. J., Norell, M. A., Yerby, S. A. & Brochu, C. A. 2004. Gigantism and comparative lifehistory parameters of tyrannosaurid dinosaurs. *Nature*, 430:772–775.

6. Tsuihiji, T., Watabe, M., Tsogtbaatar, K., Tsubamoto, T., Barsbold, R., Suzuki, S., Lee, A., Ridgely, R., Kawahara, Y. & Witmer, L. 2011. Cranial osteology of a juvenile specimen of *Tarbosaurus bataar* (Theropoda, Tyrannosauridae) from the Nemegt Formation (Upper Cretaceous) of Bugin Tsav, Mongolia. *Journal of Vertebrate Paleontology*, 31:497–517.

7. Ibid.

8. Schweitzer, M. H., Wittmeyer, J. L., Horner, J. R. & Toporski, J. K. 2005. Soft tissue vessels and cellular preservation in *Tyrannosaurus rex*. *Science*, 307:1952–1955.

9. Hone, D. W. E., Naish, D. & Cuthill, I. C. 2012. Does mutual sexual selection explain the evolution of head crests in pterosaurs and dinosaurs? *Lethaia*, 45:139–156.
10. O'Gorman, E. & Hone, D. W. E. 2012. Body size distribution of the dinosaurs. *PLOS ONE*, 7: e51925.
11. Zheng, X. T., You, H. L., Xu, X. & Dong, Z. M. 2009. An Early Cretaceous heterodontosaurid dinosaur with filamentous integumentary structures. *Nature*, 458:333–336.
12. Mallison, H. 2011. Defence capabilities of *Kentrosaurus aethiopicus Hennig*, 1915. *Palaeontologia Electronica*, 14:1–25.
13. Weishampel, D. B. 1981. Acoustic analyses of potential vocalization in lambeosaurine dinosaurs (Reptilia: Ornithischia). *Paleobiology*, 7:252–261.
14. Farke, A. A., Wolff, E. D. S. & Tanke, D. H. 2009. Evidence of combat in Triceratops. *PLOS ONE*, 4:e4252.
15. Wedel, M. J. 2009. Evidence for bird like air sacs in saurischian dinosaurs. *Journal of Experimental Zoology Part A: Ecological Genetics and Physiology*, 311:611–628.
16. Lee,Y. N., Barsbold, R., Currie, P. J., Kobayashi,Y., Lee, H. J., Godefroit, P., Escuillié, F. & Tsogtbaatar, C. 2014. Resolving the long-standing enigmas of a giant ornithomimosaur *Deinocheirus mirificus*. *Nature*, 515:257–260.
17. Lautenschlager, S. 2014. Morphological and functional diversity in therizinosaur claws and the implications for theropod claw evolution. *Proceedings of the Royal Society of London B*, 281, 20140497.
18. Hu, Y., Meng, J., Wang, Y. & Li, C. 2005. Large Mesozoic mammals fed on young dinosaurs. *Nature*, 433:149–152.
19. Happ, J. 2008. An analysis of predator-prey behaviour in a head-to-head encounter between *Tyrannosaurus rex* and *Triceratops*. In: Tyrannosaurus rex: *the Tyrant King*, Larson, P., Carpenter, K. (eds), Indiana University Press, Bloomington, pp. 355–370.
20. Rayfield, E. J. 2005. Aspects of comparative cranial mechanics in the theropod dinosaurs *Coelophysis, Allosaurus* and *Tyrannosaurus*. *Zoological Journal of the Linnean Society*, 144:309–316.

21. Dal Sasso, C. D., Maganuco, S., Buffetaut, E. & Mendez, M. A. 2005. New information on the skull of the enigmatic theropod *Spinosaurus*, with remarks on its size and affinities. *Journal of Vertebrate Paleontology*, 25:888–896.

22. Habib, M. B. 2008. Comparative evidence for quadrupedal launch in pterosaurs. *Zitteliana*, 159–166.

23. Clark *et al.* 1999. An oviraptorid skeleton from the Late Cretaceous of Ukhaa Tolgood.

24. Roach, B. T. & Brinkman, D. L. 2007. A re-evaluation of cooperative pack hunting and gregariousness in *Deinonychus antirrhopus* and other non-avian theropod dinosaurs. *Bulletin of the Peabody Museum of Natural History*, 48:103–138.

25. Stevens, K. A. 2006. Binocular vision in theropod dinosaurs. *Journal of Vertebrate Paleontology*, 26:321–330.

26. Erickson, G. M., Van Kirk, S. D., Su, J., Levenston, M. E., Caler, W. E. & Carter, D. R. 1996. Bite-force estimation for *Tyrannosaurus rex* from tooth-marked bones. *Nature*, 382:706–708.

27. Hone, D. W. E. & Watabe, M. 2010. New information on the feeding behaviour of tyrannosaurs. *Acta Palaeontologica Polonica*, 55:627–634.

28. DePalma, R. A., Burnham, D. A., Martin, L. D., Rothschild, B. M. & Larson, P. L. 2013. Physical evidence of predatory behaviour in *Tyrannosaurus rex*. *Proceedings of the National Academy of Sciences*, 110:12560–12564.

29. Fiorillo, A. R. 1991. Prey bone utilization by predatory dinosaurs. *Palaeogeography, Palaeoclimatology, Palaeoecology*, 88:157–166.

30. Chin, K., Tokaryk, T. T., Erickson, G. M. & Calk, L. C. 1998. A king-sized theropod coprolite. *Nature*, 393:680–682.

31. Persons, W. S. & Currie, P. J. 2014. Duckbills on the run: the cursorial abilities of hadrosaurs and implications for tyrannosaur-avoidance strategies. In: *Hadrosaurs*, Eberth, D. A., Evans, D. C. (eds), Indiana University Press, Bloomington, pp. 449–458.

32. Snively, E., Russell, A. P., Powell, G. L., Theodor, J. M. & Ryan, M. J. 2014. The role of the neck in the feeding behaviour of the Tyrannosauridae: inference based on kinematics and muscle function of extant avians. *Journal of Zoology*, 292:290–303.

33. Currie, P. J. 1998. Possible evidence of gregarious behaviour in tyrannosaurids. *Gaia*, 15:271–277.

34. Tanke, D. H. & Currie, P. J. 1998. Head-biting behaviour in theropod dinosaurs: paleopathological evidence. *Gaia*, 15:167–184.

35. Currie 1998. Possible evidence of gregarious behaviour.

36. Schmitz, L. & Motani, R. 2011. Nocturnality in dinosaurs inferred from scleral ring and orbit morphology. *Science*, 332:705–708.

## Moving Forwards

1. Brusatte, S. L., Norell, M. A., Carr, T. D., Erickson, G. M., Hutchinson, J. R., Balanoff, A. M., Bever, G. S., Choiniere, J. N., Makovicky, P. J. & Xu, X. 2010. Tyrannosaur paleobiology: new research on ancient exemplar organisms. *Science*, 329:1481–1485.

2. Holtz, T. R. 2008. A critical reappraisal of the obligate scavenging hypothesis for *Tyrannosaurus rex* and other tyrant dinosaurs. In: Tyrannosaurus rex: *the Tyrant King,* Larson, P., Carpenter, K. (eds), Indiana University Press, Bloomington, pp. 371–396.

3. Gilmore, C. W. 1946. A new carnivorous dinosaur from the Lance Formation of Montana. *Smithsonian Miscellaneous Collections*, 106:1–19.

4. Bakker, R. T., Williams, M. & Currie, P. J. 1988. *Nanotyrannus*, a new genus of pygmy tyrannosaur, from the latest Cretaceous of Montana. *Hunteria*, 1:1–30.

5. Carr, T. D. 1999. Craniofacial ontogeny in Tyrannosauridae (Dinosauria, Coelurosauria). *Journal of Vertebrate Paleontology*, 19:497–520.

6. Tsuihiji, T., Watabe, M., Tsogtbaatar, K., Tsubamoto, T., Barsbold, R., Suzuki, S., Lee, A., Ridgely, R., Kawahara, Y. & Witmer, L. 2011. Cranial osteology of a juvenile specimen of *Tarbosaurus bataar* (Theropoda,Tyrannosauridae) from the Nemegt Formation (Upper Cretaceous) of Bugin Tsav, Mongolia. *Journal of Vertebrate Paleontology*, 31:497–517.

7. Horner, J. R. & Lessem, D. 1993. *The Complete* T. rex: *How Stunning New Discoveries Are Changing Our Understanding of the World's Most Famous Dinosaur.* Simon & Schuster.

8. Larson, P. 2008. Variation and sexual dimorphism in *Tyrannosaurus rex.* In: Tyrannosaurus rex: *the Tyrant King*, Larson, P., Carpenter, K. (eds), Indiana University Press, Bloomington, pp. 103–130.

9. Schweitzer, M. H., Suo, Z., Avci, R., Asara, J. M., Allen, M. A., Arce, F. T. & Horner, J. R. 2007. Analyses of soft tissue from *Tyrannosaurus rex* suggest the presence of protein. *Science*, 361:277–280.

# Further Reading

The scientific literature can be tough to wade through and heavy reading; even on a good day, there's only so much about denticle counts and femur-tibia ratios I can stomach. As noted in the introduction, I have tried to keep the number of citations down (a work this size in the formal scientific literature would probably have hundreds of references listed at this point), and to stick to those that are most relevant or significant (or are ones that I wrote). There is simply not enough room to include everything, and certainly there are more references that could have been added. The earliest, most recent and most important papers on a given subject tend to get preference here, but this doesn't mean others are not important.

Not all of these references are easily accessible, but increasing numbers of papers are available online, and those publishing their own research are generally happy to supply copies on request: they *want* you to read their work. The Google Scholar and PubMed online searches can help turn these up, then you just have to read them. Diving into the literature can be difficult, so here are five works that should help you to tackle the more in-depth/tedious stuff in the order in which it is best approached, followed by some additional sources that helped in the writing of this book and provide a great deal more scientific knowledge of the tyrannosaurs.

1. *Encyclopedia of Dinosaurs.* 2007. Thomas R. Holtz Jnr, illustrated by Luis Rey, Random House.
Modern classic on dinosaurs and a rare book for which the term 'suitable for all ages' applies.
2. *The Great Dinosaur Discoveries.* 2009. Darren Naish. A & C Black.
Covers more of the history of dinosaur palaeontology and focuses on key finds, and how they improved and changed our understanding of dinosaurs.

3. *The Complete Dinosaur*. 2012. Michael Brett-Surman, Thomas R. Holtz Jnr & James O. Farlow. Indiana University Press. Covers pretty much everything: excavation, history, research practices, behaviour, anatomy, and even artwork and museum displays.

4. *Dinosaur Paleobiology*. 2012. Stephen L. Brusatte. Wiley-Blackwell. Recent summary of dinosaur basics and research aimed at undergraduates and early postgraduates. Helps back up the former works and acts as a bridge into technical papers.

5. *Dinosauria* (2nd edn). 2004. David D. Weishampel, Peter Dodson & Halszka Osmólska. University of California Press. Technical work covering the origin, evolution, diversity and ecology of the dinosaurs (especially their anatomy and systematic relationships). Essentially *the* major 'bible' of dinosaurian research, though it is starting to date now; a new edition is in the works but probably won't be out for another year or so.

Abler, W. L. 2001. A kerf-and-drill model of tyrannosaur tooth serrations. In: *Mesozoic Vertebrate Life*, Tanke, D. H., Carpenter, K., Skrepnick, M. W. (eds), Indiana University Press, Bloomington, pp. 84–89.

Barrett, P. M. & Rayfield, E. J. 2006. Ecological and evolutionary implications of dinosaur feeding behaviour. *Trends in Ecology & Evolution*, 21:217–224.

Bates, K. T. & Falkingham, P. L. 2012. Estimating maximum bite performance in *Tyrannosaurus rex* using multi-body dynamics. *Biology Letters*, 8:660–664.

Benson, R. B. J. 2008. New information on *Stokesosaurus*, a tyrannosauroid (Dinosauria: Theropoda) from North America and the United Kingdom. *Journal of Vertebrate Paleontology*, 28:732–750.

Breithaup, B. H., Southwell, E. H. & Matthews, N. A. 2006. *Dynamosaurus imperiosus* and the earliest discoveries of *Tyrannosaurus rex* in Wyoming and the West. *New Mexico Museum of Natural History and Science Bulletin*, 35:257–258.

Brochu, C. A. 2003. Osteology of *Tyrannosaurus rex*: insights from a nearly complete skeleton and high-resolution computed tomographic analysis of the skull. *Society of Vertebrate Paleontology Memoir*, 7:1–138.

Brusatte, S. & Benson, R. B. J. 2013. The systematics of Late Jurassic tyrannosauroids (Dinosauria: Theropoda) from Europe and North America. *Acta Palaeontologica Polonica*, 58:47–54.

Brusatte, S. L., Carr, T. D. & Norell, M.A. 2012. The osteology of *Alioramus*, a gracile and long-snouted tyrannosaurid (Dinosauria: Theropoda) from the Late Cretaceous of Mongolia. *Bulletin of the American Museum of Natural History*, 366:1–197.

Buckland, W. 1824. Notice on the *Megalosaurus* or great Fossil Lizard of Stonesfield. *Transactions of the Geological Society of London*, 2:390–396.

Cadbury, D. 2000. *The Dinosaur Hunters*. Fourth Estate, London.

Carpenter, K. 1992. Tyrannosaurids (Dinosauria) of Asia and North America. In: *Aspects of Nonmarine Cretaceous Geology*, Mateer, N., Chen, P. J. (eds) China Ocean Press, Beijing, pp. 250–268.

Carr, T. D. & Williamson, T. E. 2010. *Bistahieversor sealeyi*, gen. et sp. nov., a new tyrannosauroid from New Mexico and the origin of deep snouts in Tyrannosauroidea. *Journal of Vertebrate Paleontology*, 30:1–16.

Carr, T. D., Williamson, T. E. & Schwimmer, D. R. 2005. A new genus and species of tyrannosauroid from the Late Cretaceous (middle Campanian) Demopolis Formation of Alabama. *Journal of Vertebrate Paleontology*, 25:119–143.

Colbert, E. H. 1968. *Men and Dinosaurs: The Search in Field and Laboratory*. Evans Brothers, Limited.

Cooper, L. N., Lee, A. H., Taper, M. L. & Horner, J. R. 2008. Relative growth rates of predator and prey dinosaurs reflect effects of predation. *Proceedings of the Royal Society B*, 275:2609–2615.

Currie, P. J. 2000. Theropods from the Cretaceous of Mongolia. In: *The Age of Dinosaurs in Russia and Mongolia*, Benton, M. J., Shishkin, M. A., Unwin, D. M., Kurochkin, E. M. (eds), Cambridge University Press, Cambridge, pp. 434–455.

Currie, P. J., Trexler, D., Koppelhus, E. B., Wicks, K. & Murphy, N. 2005. An unusual multi-individual tyrannosaurid bone bed in the Two Medicine Formation (Late Cretaceous, Campanian) of Montana (USA). In: *The Carnivorous Dinosaurs*, Carpenter, K. (ed.) Indiana University Press, Bloomington, pp. 313–324.

Erickson, G. M., Currie, P. J., Inouye, B. D. & Winn, A. A. 2006. Tyrannosaur life tables: an example of non-avian dinosaur population biology. *Science*, 313:213–217.

Farlow, J. M., Brinkman, D. L., Abler, W. L. & Currie, P. J. 1991. Size, shape, and serration density of theropod dinosaur lateral teeth. *Modern Geology*, 16:161–198.

Fowler, D. W., Woodward, H. N., Freedman, E. A., Larson, P. L. & Horner, J. R. 2011. Reanalysis of '*Raptorex kriegsteini*': a juvenile tyrannosaurid dinosaur from Mongolia. *PLOS ONE*, 6:e21376.

Holliday, C. M. 2009. New insights into dinosaur jaw muscle anatomy. *The Anatomical Record*, 292:1246–1265.

Holtz, T. R. 2001. The phylogeny and taxonomy of the Tyrannosauridae. In: *Mesozoic Vertebrate Life*. Tanke, D. H., Carpenter, K. & Skrepnick, M. W. (eds), Indiana University Press, Bloomington, pp. 64–83.

Holtz, T. R. 2003. Dinosaur predation: evidence and ecomorphology. In: *Predator-Prey Interactions in the Fossil Record*. Kelly, P. H., Kowalewski, M., Hansen, T. A. (eds), Kluwer Academic/Plenum Publishers, New York, pp. 325–340.

Hone, D. W. E. & Rauhut, O. W. M. 2010. Feeding behaviour and bone utilization by theropod dinosaurs. *Lethaia*, 43:232–244.

Hone, D. W. E. & Tanke, D. H. 2015. Pre- and post-mortem tyrannosaurid bite marks on the remains of *Daspletosaurus* (Tyrannosaurinae: Theropoda) from Dinosaur Provincial Park, Alberta, Canada. *PeerJ*, 3:e885.

Horner, J. R. 1994. Steak knives, beady eyes, and tiny little arms (a portrait of *Tyrannosaurus* as a scavenger). *The Paleontological Society Special Publication*, 7:157–164.

Hurum, J. H. & Sabath, K. 2003. Giant theropod dinosaurs from Asia and North America: Skulls of *Tarbosaurus bataar* and *Tyrannosaurus rex* compared. *Acta Palaeontologica Polonica*, 48:161–190.

Hutchinson, J. R. & Garcia, M. 2002. *Tyrannosaurus* was not a fast runner. *Nature*, 415:1018–1021

Hutt, S., Naish, D., Martill, D. M., Barker, M. J. & Newbery, P. 2001. A preliminary account of a new tyrannosauroid theropod from the Wessex Formation (Early Cretaceous) of southern England. *Cretaceous Research*, 22:227–242.

Knell, R., Naish, D., Tompkins, J. L. & Hone, D. W. E. 2013. Sexual selection in prehistoric animals: detection and implications. *Trends in Ecology and Evolution*, 28:38–47.

Lambe, L. M. 1917. The Cretaceous theropodus dinosaur *Gorgosaurus*. *Government Printing Bureau*, vol. 100.

Larson, P. 2008. Variation and sexual dimorphism in *Tyrannosaurus rex*. In: Tyrannosaurus rex, *The Tyrant King*. Larson, P., Carpenter, K. (eds), Indiana University Press, Bloomington, pp. 102–128.

Maleev, E. A. 1955. New carnivorous dinosaurs from the Upper Cretaceous of Mongolia. *Doklady Akademii Nauk SSSR*, 104:779–782.

Molnar, R. E. 1991. The cranial morphology of *Tyrannosaurus rex*. *Palaeontographica A*, 217:137–176.

Osborn, H. F. 1906. *Tyrannosaurus*, Upper Cretaceous carnivorous dinosaur (second communication). *Bulletin of the American Museum of Natural History*, 22: 281–296.

Owen, R. 1842. Report on British Fossil Reptiles. Part II. Report of the Eleventh Meeting of the British Association for the Advancement of Science; Held at Plymouth in July 1841. London, pp. 60–204.

Rauhut, O. W. M. 2003. A tyrannosauroid dinosaur from the Upper Jurassic of Portugal. *Palaeontology*, 46:903–910

Ricklefs, R. E. 2007. Tyrannosaur ageing. *Biology Letters*, 3:214–217.

Russell, D. A. 1970. Tyrannosaurs from the Late Cretaceous of western Canada. *National Museum of Natural Sciences Publications in Paleontology*, 1:1–34.

Sampson, S. D. 2009. *Dinosaur Odyssey: Fossil Threads in the Web of Life*. University of California Press.

Sereno, P. C., Tan, L., Brusatte, S. L., Kriegstein, H. J., Zhao, X. & Cloward, K. 2009. Tyrannosaurid skeletal design first evolved at small body size. *Science*, 326:418–422.

Snively, E., Henderson, D. M. & Phillips, D. S. 2006. Fused and vaulted nasals of tyrannosaurid dinosaurs: implications for cranial strength and feeding mechanics. *Acta Palaeontologica Polonica*, 51:435–454.

Varricchio, D. J. 2001. Gut contents from a Cretaceous tyrannosaurid: implications for theropod dinosaur digestive tracts. *Journal of Paleontology*, 75:401–406.

Witmer, L. M. & Ridgely, R. C. 2009. New insights into the brain, braincase, and ear region of tyrannosaurs (Dinosauria, Theropoda), with implications for sensory organization and behaviour. *Anatomical Record*, 292:1266–1296.

Witmer, L. M. & Ridgely, R. C. 2010. The Cleveland tyrannosaur skull (*Nanotyrannus* or *Tyrannosaurus*): new findings based on CT scanning, with special reference to the braincase. *Kirtlandia*, 51:61–81.

Zelenitsky, D. K., Therrien, F. & Kobayashi, Y. 2009. Olfactory acuity in theropods: palaeobiological and evolutionary implications. *Proceedings of the Royal Society B: Biological Sciences*, 276:667–673.

# Museums and Online Sources

Following on from the traditional references of scientific papers, associated books and guides that might tempt dinosaur enthusiasts, here is a list of alternate places (both physical and digital) that are superb for finding out more information.

## Museums

A great many museums around the world house at least a cast (that is, a copy taken from a mould of an original bone or skeleton) of a *Tyrannosaurus* skull, or even a whole skeleton. Seeing real specimens (beyond the odd tooth or bone) can be a bit of a challenge, however. Despite the wide range of the tyrannosaurs, as discussed earlier the vast majority of fossils comes from North America and eastern Asia. They are both important scientifically and valuable, and consequently most specimens are kept in museums and many are not on display. Specimens can be at risk when exposed to lights and the changing environmental conditions of a museum exhibit (not to mention the public), so it is not always safe to display them even if researchers would like to.

Here, then, is a partial list of places that currently have tyrannosaurs on show (this situation does vary, with more or fewer specimens being on show at different times). Collections with especially large numbers of specimens or important ones are in bold. If you want to see a cast of a tyrannosaur, any of the great museums here (and many more besides) are a great start.

## Europe

UK: Natural History Museum, London; Oxford University Museum of Natural History, Oxford; National Museum of Scotland, Edinburgh.
France: National Museum of Natural History, Paris.
Germany: Senckenberg Museum, Frankfurt.

## Asia and Australia

Japan: **Fukui Prefectural Dinosaur Museum**, Fukui; National Museum of Nature and Science, Tokyo.
China: Palaeozoological Museum of China, Beijing.
Australia: National Dinosaur Museum, Canberra.

## North America

Canada: **Royal Tyrrell Museum of Palaeontology**, Alberta; Royal Ontario Museum, Ontario.
USA: **Field Museum of Natural History**, Chicago; **Natural History Museum of Los Angeles County**, Los Angeles; **Smithsonian National Museum of Natural History**, Washington; **Carnegie Museum of Natural History**, Pittsburgh; **Museum of the Rockies**, Montana.
Mexico: Museo del Desierto, Coahuila.

## Websites and blogs

The web has revolutionised how scientists communicate both with each other and with the public. However, it has also made it easier than ever for huge amounts of misinformation to be shared by people, and promoted as being accurate or true; wading through it all can be difficult without knowing what is correct or out of date. Here are a few tyrannosaur and general dinosaur websites and blogs that may be of interest, and that are written by those with more than a passing familiarity with the relevant subjects.

davehone.co.uk
My own research pages. Yes, a bit of self-promotion, but also many of my papers are available here, including various ones that involve tyrannosaurs.
https://archosaurmusings.wordpress.com
The Archosaur Musings. My first major blog covering dinosaurs, pterosaurs and science communication issues. Check out the huge series on *Gorgosaurus* especially.
www.theguardian.com/science/lost-worlds

The Lost Worlds. My second major blog for the *Guardian* online. Less technical than the Musings.

http://askabiologist.org.uk

Ask a Biologist. A question-and-answer site for all ages. Covers all of biology, but as I set this up I inevitably recruited quite a few palaeontologists, so there's a superb bank of people who can tackle any tyrannosaur-related queries in particular.

www.skeletaldrawing.com

Skeletal Drawing. Scott Hartman's website and blog on dinosaur anatomy and art.

http://tyrannosauroideacentral.blogspot.co.uk

Tyrannosauroidea Central. Tyrannosaur specialist Thomas Carr's blog about the tyrannosaurs.

https://dinosaurpalaeo.wordpress.com

Dinosaurpalaeo. Dinosaur blog by palaeontologist Heinrich Mallison that also features a fair number of living species.

http://svpow.com

SV-POW! Site devoted to the sauropod dinosaurs, and more specifically their vertebrae.

https://luisvrey.wordpress.com

Luis Rey's art. Artworks by award-winning palaeartist Luis Rey. Brilliant and bold images.

http://chasmosaurs.blogspot.co.uk

Love in the Time of Chasmosaurus. Light-hearted look at dinosaurs, and especially old artworks and ideas.

http://blogs.scientificamerican.com/tetrapod-zoology

Tetrapod Zoology. One of the best biology blogs online, covering a lot of dinosaur material as it's by palaeontologist (and *Eotyrannus* describer) Darren Naish.

# Acknowledgements

There are numerous palaeontologists (including researchers, fieldworkers, preparators, artists and others) who have contributed major works on tyrannosaurs and who I have had the fortune to work with, or who have given me their time and assistance in discussing these amazing animals and their history. All of this has fed through one way or another into this book, and therefore I must thank Xu Xing, Darren Naish, Darren Tanke, Phil Currie, John Hutchinson, Mahito Watabe, Scott Persons, Larry Witmer, Oliver Rauhut, Emily Rayfield, Mark Loewen, Don Henderson, Steve Brusatte, Mike Habib, Taka Tsuihiji and Corwin Sullivan for their help, assistance, friendship and knowledge. I am also indebted to all those who have published their scientific works on tyrannosaurs, without which this book would be far shorter.

Special thanks go to Tom Holtz, who has put up with many questions from me on the tyrannosaurs, and was good enough to read through the entire manuscript to offer his thoughts and check for errors. This both helped correct various mistakes I had made, and means that any others that are still there I can blame him for not spotting.

Similarly, I thank Charlie Dobbie, Marc Vincent and Heledd Wynn for reading through the book and helping to identify areas that needed clarification for the non-tyrannosaur expert. Their time was much appreciated.

Scott Hartman has long been someone I can rely on when it comes to the graphical representations of long-extinct animals, and I was delighted that he agreed to illustrate this book. It would not be the same without his contribution. I must also thank Xu Xing, Larry Witmer, John Hutchinson, Jordan Mallon, Andy Farke, Lü Junchang, Pete Falkingham, Phil Currie and Mark Loewen, and the Carnegie Museum, The Royal Tyrrell Museum, The Institute of Vertebrate Palaeontology and Palaeoanthropology, Dinosaur Isle, The Los Angeles County Museum, the Royal Saskatchewan Museum, The New Mexico Museum for Natural History and Science, the Mongolian Academy of Sciences and the Hayashibara Museum for contributing images used in this book, or providing permission to publish photographs of specimens in their care.

At Bloomsbury I'd like to thank Jim Martin for giving me the opportunity to write my book and allowing my ideas to flow. My thanks to Anna MacDiarmid for her work on this and setting the central spread of pictures and to Krystyna Mayer for her work proofing the text. This book would be very different without them.

# Index

abelisaurs 153–54, 203
aetosaurs 38
air sacs 32, 42, 105–107
albertosaurines 13, 46, 47–48, 54, 70,
    78, 80, 82, 113, 152, 155, 203–204,
    219, 252, 266–67, 268
*Albertosaurus* 28–29, 62, 70, 78, 167,
    201, 213, 237, 241–43
*Alioramus* 12, 30, 54, 70, 78, 82, 93–94,
    151, 152, 155, 161, 174, 210, 213
    *Alioramus altai* 52
allosauroids 45, 53, 69, 148
allosaurs 155, 202–204, 213, 216,
    234, 252
    *Allosaurus* 27, 43, 67, 95, 203
alvarezsaurs 75, 81, 190, 216
ancestral animals 147–50
Andrews, Roy Chapman 30
ankylosaurs 40, 83, 178, 181–83, 190,
    204, 225, 227, 230, 265
antorbital fenestra 17
*Apatosaurus* 27, 41, 66, 102, 188
*Archaeopteryx* 32, 83, 123
archosaurs 38–39, 87, 92, 94, 96, 107,
    124, 138, 140, 161, 175, 207, 209
arctometatarsus 17, 47, 150, 191–92
*Argentinosaurus* 190
arms 153–55, 232–33
atlas vertebra 18
*Aviatyrannis* 54, 252
azhdarchids 202, 207–208, 213, 215

Bakker, Bob 138, 253, 256
*Baryonyx* 204–206
behaviour 237–38
    evidence for social behaviour 241–45
    getting up and lying down 245–46
    solitary lifestyles 238–41
    speculation on possible
        behaviours 246–48
*Beipiaosaurus* 194
Benson, Roger 61–62
Bergmann's Rule 155–56, 266
birds 15, 20, 26, 105, 106–107, 157, 246
    dinosaur descent 32, 42, 67, 140
    eggs 161–64
    feathers 123–28
    fingers 113–14

nests 164–68
pair bonding 247–48
*Bistahieversor* 69, 70, 151, 213
bite patterns 233–35
body 99–100
    belly bones 100–104
    insides 105–109
bone structure 17–20
*Brachiosaurus* 41, 188, 264
*Brontosaurus* 188
Brown, Barnum 27, 29, 241
Brusatte, Steve 51
Buckland, William 25

*Caenorhabditis elegans* 251
Cambrian 77
carbon dating 73–74
carcharodontosaurs 203–204, 213, 216
carnivory 150–53, 178
carpals 18
Carr, Thomas 254
caudal vertebrae 18
*Caudipteryx* 102
ceratopsians 81, 83, 155, 161, 184,
    186–89, 198, 204, 221, 224–25, 227,
    229–30, 232–33, 239, 242, 255, 273
ceratosauroids 53
ceratosaurs 114, 202–203, 219
    *Ceratosaurus* 44, 205
champsosaurs 225, 273
    *Champsosaurus* 195
chevrons 18
*Chingkankousaurus* 51, 53, 55
clades 11, 35
    phylogeny of the major
        tyrannosaur clades 13–14, 54
cladistics 35–36
claws 131–32
cloaca 107–108
coelurosaurs 43, 67, 73, 124
cold-blooded animals 134
colour 268–70
competition 172–75
competitors 199–202, 212–17
    azhdarchids 207–208
    ceratosaurs, megalosaurs and
        allosaurs 202–204
    crocodilians 206–207, 208–209

dromaeosaurs   211—12
  evolution   202
  oviraptorosaurs   210
  pterosaurs   206, 207–208
  spinosaurs   204–206
  theropods   209–10
  troodontids   210–11
compsognathids   43, 67, 107, 124–25, 148
computer models   270–71
continental drift   14–15, 81
Cope, Edward   26–27, 28
coprolites   107–108, 224, 226, 235, 277
coracoid   18
cranium   17
crests   172–74
Cretaceous   15, 28, 41, 69, 78, 145,
  153, 177, 179, 181, 182, 189–90,
  191, 200, 206, 213, 254, 269, 276
Cretaceous Terrestrial Revolution   83
  faunal changeover   82–84, 204
crocodiles   15, 25, 37, 66, 93, 108,
  123, 127, 129, 168, 171, 194, 214,
  225, 227, 247
crocodilians   38, 94, 105, 130, 137–38,
  142, 161, 164–66, 185, 195,
  205–206, 208–209, 213, 215, 219,
  234, 238, 242, 246, 273
Currie, Phil   242–43
cycads   15, 83, 177

Darwin, Charles On the Origin of
  Species   35
Daspletosaurus   29, 54, 62, 70, 152,
  167, 213, 226
Deinocheirus   192
Deinodon   28, 51, 53
deinonychosaurs   67
Deinonychus   205, 212
Deinosuchus   208–209
dentaries   17
Dickens, Charles Bleak House   26
Dilong   33, 54, 62, 78, 124, 139, 148,
  149, 195, 202, 213, 266
dinosaurs   23–25
  archosaurs   38–41
  dinosaur renaissance   31–34
  extinction   273–76
  nineteenth century   25–29
  reptiles   36–38
  theropods   41–43
  twentieth century   29–31
  tyrants   43–49
  what are the dinosaurs?   35–36

Diplodocus   12, 27, 41, 66, 180, 188
distribution   79–85
dorsal ribs   18
dorsal vertebrae   18
Doyle, Conan The Lost World   26
Drepanosaurus   38
dromaeosaurs   67, 81, 190, 205, 209,
  211–13, 215, 220, 254
Drosophilus melanogaster   251
Dryosaurus   264
Dryptosaurus   54, 80
Dynamosaurus   28, 51

Early Cretaceous   78, 80, 82, 183, 187,
  203, 204
Early Jurassic   75, 78, 267
ectothermy   135–41, 200, 221
Edmontosaurus   185
eggs   30, 59, 108, 161–64, 175, 210,
  262–63
End Cretaceous   15, 199, 213, 273–76
endothermy   135–41
Eotyrannus   33, 54, 80, 147–48, 151, 245
evolution   147–50
  head size and shape   150–53
  increasing body size   155–57
  short arms   153–55
extinction   273–76

feathers   32, 42, 123–28, 141, 149,
  193–94
  colour   269–70
  future finds   265–68
feeding   219
  bite patterns   233–35
  finding a meal   227–29
  killing techniques   229–33
  predation and scavenging   220–24
  who is on the menu?   224–27
feet   154–55
femur   19
ferns   15, 83, 177
fibula   15
fossilisation   76–77
Fukuiraptor   69
future developments   261–62
  animals of a feather   265–68
  colour   268–70
  computer models   270–71
  things to come   262–65

gastralia   18
genus (genera)   11–13

geographic context   14–16
getting up   245–46
Gilmore, Charles   253
*Gorgosaurus*   29, 46, 54, 62, 70, 78,
    102, 125, 201, 213, 253, 254, 265
grasses   15
growth   139–40, 166–72
*Guanlong*   16, 45, 47, 54, 62, 70, 75, 78, 80,
    113, 116, 147, 151, 152, 169, 173, 245

hadrosauroids   183–84
hadrosaurs   66, 81–83, 119, 143, 161,
    174, 184–86, 187, 198, 201, 204,
    208, 239, 242, 244, 273
    hadrosaurs as prey   223–27, 229–31,
    233–35
    *Hadrosaurus*   27, 40
Hayden, Ferdinand   27
heterodontosaurs   178–80
heterothermy   135–36
Holtz, Tom   252
homeothermy   135–41
Horner, Jack   220
horsetails   15, 177
humerus   18

icthyosaurs   15, 25, 38
*Iguanodon*   26–27, 40, 181, 183–84, 206
ilium   19
insects   15
internal organs   105–109
intromittent organs   108–109, 175
ischium   19
Island Rule   156

Jurassic   15, 41, 82, 177, 179, 190, 206,
    212–13, 269
*Jurassic Park*   23, 210
juveniles   56, 60, 168–72, 265
    as prey   226–27, 229
    growth rate   139–40, 166–68
    *Nanotyrannus*   253–56

*Kileskus*   54, 75, 78
killing techniques   229–33
*King Kong*   23

Lakes, Arthur   28
Late Cretaceous   15, 27, 77, 78, 80,
    83, 182, 184, 186–87, 202, 203,
    204, 207, 209, 210, 243, 266
Late Jurassic   83, 181, 267
Late Triassic   15

Leidy, Joseph   27–28, 133
lifespans   142
limbs   111–15
    gait   118–21
    legs   116–18
lips   129–30
lizards   15, 26, 31, 38, 123, 130, 134,
    137, 138, 157, 161, 168
Loewen, Mark   69, 71, 82
lying down   245–46
*Lythronax*   33, 54, 69, 78, 151, 213, 252

Maleev, Evgeny   30
mammals   15, 20, 26, 134–39, 141–42,
    157, 168, 194–95
mandibles   17
maniraptorans   83, 108, 179
Mantell, Gideon   26–27, 31, 133
Marsh, Charles   26–27, 28
mass extinction event   273–76
mating   108–109, 174–75
maxilla   17
medullary bone   59, 171, 257–58
megalosaurs   202–204, 213, 216,
    234, 252
    *Megalosaurus*   25–28, 203
*Megaraptor*   69
Mesozoic   32, 38, 77, 82, 138, 139,
    175, 177, 190, 194, 199, 202, 206,
    208, 212, 273
metacarpals   18
metatarsals   15
meteoric impact   274–75
Middle Cretaceous   186–87
Middle Jurassic   15, 75, 77, 78, 83,
    181, 186, 240, 267, 278
*Monolophosaurus*   45
muscle distribution   130–31

*Nanotyrannus*   51, 60, 253–56
*Nanuqsaurus*   33, 54, 155–56, 252,
    261–62, 264, 266
nares   17
necks   99–100, 188–89, 230
nests   32, 164–68
Novas, Fernando   69

orbits   17
ornithischians   39–42, 66, 124, 139–41,
    177–79, 187, 189–90, 211–12, 230
ornithomimosaurs   44, 67, 148, 154,
    178, 191–93, 196, 198, 216, 225,
    230, 266

Osborn, Henry Fairfield   28
oviraptorosaurs   47, 81, 102, 124,
      143, 155, 161, 190, 192, 210,
      225, 247
Owen, Sir Richard   25–27, 32, 133

pachycephalosaurs   187–88
pair bonding   247–48
Parasaurolophus   185
parental care   164–68
Pelecanimimus   191
phalanges   18, 19
phylogenies   63–65
      major groups in the theropods   66
      tyrannosaur clades   13–14, 54, 70
physiology   133
      hot- or cold-blooded?   133–37
      size   142–46
      temperature control   140–42
      upright posture   137–40
Pinacosaurus   182
placodonts   38
plesiosaurs   25, 38, 273
predation   220–24
predator traps   241–42
premaxilla   17
prey   177–78
      animals other than dinosaurs
         194–95, 265
      ankylosaurs   181–83
      basal hadrosauroids   183–84
      ceratopsians   186–87
      hadrosaurs   184–86
      heterodontosaurs   178–80
      ornithomimosaurs   191–92
      pachycephalosaurs   187–88
      sauropods   188–90
      stegosaurs   180–81
      survival   195–98
      therizinosaurs   192–94
      theropods   190–91
proceratosaurids   13, 45, 46, 54, 67,
      70, 71, 80, 90, 151, 155, 173–74,
      209, 267–68
      Proceratosaurus   33, 44, 45, 52, 75, 78,
         80, 173, 252, 264
Psittacosaurus   186
pterosaurs   15, 38, 43, 66–67, 105,
      117, 124, 138, 140–41, 161, 178,
      195, 198, 202, 206–207, 209, 211,
      221, 225, 245, 273, 276
      azhdarchids   207–208
pubis   19

Qianzhousaurus   33, 54

radius   18
Raptorex   54, 60, 161, 169, 171,
      213, 264
Rauhut, Oliver   52
relationships   63–65
      evolutionary place of
         tyrannosaurs   65–72
Repenomamus   194–95
reproduction   161–64
      competitive features   172–75
      growth   168–72
      nests   164–68
reptiles   15, 20, 25–26, 31, 123, 139
      what is a reptile?   36–38
research   7–8, 251–53, 258–59
      a tiny tyrant?   253–56
      two Tyrannosaurus?   256–58
rhynchosaurs   38
ribs   18

sacral vertebrae   18
saurischians   39–42, 45
Saurolophus   82, 221–23
sauropodomorphs   40–41, 66, 124
sauropods   41, 66, 81, 83, 101,
      105, 140, 143, 155, 162, 175,
      177, 183–84, 188–90, 224,
      227, 273
scales   123–28
      beyond scales   128–32
scansoriopterygids   124
scapula   18
scavenging   220–24
scientific names   11–13, 28, 53–54
Scipionyx   107
sexual dimorphism   58–59
Shanshanosaurus   51
Shenzhousaurus   191
Sinocalliopteryx   148
Sinraptor   240
size   142–46, 155–57
skeleton   17–20, 170–71
skin   123–28
      beyond scales   128–32
skull   17–18, 87–90
      brain boxes   90–94
      teeth   94–97
sleep   244–45
snakes   15, 38, 105, 130, 137, 161
social behaviour   237–38
solitary lifestyles   238–41

species 11–13, 51–54
  naming a new species 60–62
  what makes a species a
    species? 55–60
spinosaurs 53, 195, 202, 204–206,
  214–15
  *Spinosaurus* 24, 203, 204, 205
stegosaurs 108, 175, 180–81, 182, 190
  *Stegosaurus* 27, 40, 180
Sternberg, George 27, 29
*Stokesosaurus* 46, 54, 80, 147, 181,
  202, 203, 263
*Styracosaurus* 186
survival 195–98

tails 57–58, 102–104, 230–31
*Tanystropheus* 38
*Tarbosaurus* 30, 53, 54, 56, 60, 61, 63,
  71, 78, 125, 143, 152, 153, 155,
  161, 169, 213, 222–23, 234, 251,
  254, 256
tarsals 15
taxonomy 35–36
teeth 94–97, 149
*Tenontosaurus* 212
*Teratophoneus* 46, 54, 80, 147, 181,
  202, 203, 263
therizinosaurs 67, 124, 143, 161, 181,
  192–93, 230
theropods 26, 32, 40–43, 83, 245
  fingers 113–14
  major groups in the theropods 66
*Tianyulong* 139, 179–80, 187
tibia 15
timescale 14–16, 73–75
  distribution 79–85
  the oldest tyrants 75–78
tortoises 15, 37, 123, 127, 128, 157,
  192, 194
Triassic 15, 38, 75, 190
*Triceratops* 12, 24, 40, 125, 173–74,
  180, 186–87, 198, 221, 227, 231,
  234, 244, 265
*Troodon* 27, 139
troodontids 44, 81, 108, 190, 209,
  210–13, 215, 220, 225, 254
tyrannosaurids 13, 44–45, 47–48,
  154, 228, 231–32, 268
tyrannosaurines 13, 47–48, 252, 268
  evolution 153, 154, 155
  feeding 219, 224, 228, 232, 234
  'Jane' 255–56
  phylogenies 69–71

tyrannosauroids 13–14, 44, 47, 153,
  154, 181, 232
  evolution 153, 154, 155
tyrannosaurs 23–25, 276–79
  bony anatomy 17–20
  illustration of fossils 9–10
  names 11–13
  relationships 13–14
  research 7–8
  timescale 14–16
  tyrant dinosaurs 14, 43–49
*Tyrannosaurus* 7, 16, 23, 28, 29, 43, 47,
  51, 62, 141, 180, 198, 203, 205
  extinction 78, 273
  feeding 219, 220–21, 223, 226,
    228, 235
  future finds 263, 265, 266, 268
  growth rates 166–68
  head size 152
  internal organs 106, 107
  *Nanotyrannus* 253–56
  phylogenies 67, 71, 82
  research 251–53, 259
  size 143, 144, 145, 267
  skull 88–89, 90
  'Sue' 24, 33, 91, 145, 146, 257, 267
  two fingers 113
  two *Tyrannosaurus*? 256–58
*Tyrannosaurus rex* 12, 23, 27, 57, 199,
  256–57
Tyrrell, Joseph 28–29

ulna 18
unguals 18–19
upright posture 137–40

*Velociraptor* 30, 67, 95, 195, 210
vertebrae 18
vulcanism 274

walking 118–21
  upright posture 137–40
warm-blooded animals 133–37

*Xenopus laevis* 251
*Xiongguanlong* 54, 80, 100, 115

*Yutyrannus* 33, 52, 54, 124–25, 139,
  151, 152, 154, 174, 213, 216, 252,
  266, 269–70

*Zhuchengtyrannus* 33, 54, 60–61, 63,
  143, 145, 213, 251, 264